REAL-TIME SYSTEMS IN MECHATRONIC APPLICATIONS

edited by

Jan Wikander
Royal Institute of Technology

Bertil Svensson
Chalmers University of Technology
and
Halmstad University

A Special Issue of
REAL-TIME SYSTEMS
The International Journal of Time-Critical Computing Systems
Volume 14, No. 3 (1998)

KLUWER ACADEMIC PUBLISHERS
Boston / Dordrecht / London

REAL-TIME SYSTEMS

The International Journal of Time-Critical Computing Systems

Volume 14, No. 3, May 1998

Special Issue on Real-Time Systems in Mechatronic Applications
Guest Editors: Jan Wikander and Bertil Svensson

Distributors for North, Central and South America:
Kluwer Academic Publishers
101 Philip Drive
Assinippi Park
Norwell, Massachusetts 02061 USA

Distributors for all other countries:
Kluwer Academic Publishers Group
Distribution Centre
Post Office Box 322
3300 AH Dordrecht, THE NETHERLANDS

ISBN 978-1-4419-5037-6   e-ISBN 978-0-585-35223-7

Library of Congress Cataloging-in-Publication Data

A C.I.P. Catalogue record for this book is available
from the Library of Congress.

Real-Time Systems, 14, 217–218 (1998)
© 1998 Kluwer Academic Publishers, Boston.

Editorial

JAN WIKANDER   jan@damek.kth.se
Department of Machine Design, Mechatronics Lab, Royal Institute of Technology, S-100 44 Stockholm, Sweden

BERTIL SVENSSON   bertil.svensson@cca.hh.se;svensson@ce.chalmers.se
Department of Computer Engineering, Chalmers University of Technology, S-41296 Gothenburg, Sweden
and Centre for Computer Systems Architecture, Halmstad University, Box 823, S-30118 Halmstad, Sweden

1. Introduction

The functionality and performance of modern machines rely heavily on integrated computer control systems. It is even so that in many machines the computer system is no longer just added on, but rather integrated to such an extent that the functionality of the machine is directly dependent on the real-time control system, such that the machine is useless if the computer system does not work properly. An example of this is the modern diesel engine for trucks and buses, which can not at all meet the environmental requirements without its control system, which in turn has such requirements and complexity that the only possibility is a computer based control system. The physical embedding and integration of the control system into the machine is another unique characteristic for a real-time mechatronic control system. Again, the diesel engine serves as a good example, typically we find the embedded controller bolted directly onto the engine block. This certainly puts specific requirements on the design and packaging of the control system.

Given this, and having in mind that many of the functions and applications that are realized in these systems are critical in terms of safety, e.g. spin-control in a car or control of medical equipment, it is clear that engineering of real-time systems for mechatronic applications requires a lot of attention.

The initial ambition with this special issue on real-time systems in mechatronic applications was to put together a set of interesting contributions helping to improve real-time systems engineering in mechatronic applications. Papers covering methods, tools, strategies, components and industrial case studies were looked for. We are now pleased to present the result of this in terms of six papers with a reasonably broad coverage of the area.

2. Contributions

The contributions to this special issue cover the whole development cycle associated with the design and implementation of real-time mechatronic applications. That is, specification and modeling; design and programming; rapid prototyping; implementation; and finally some applications.

The first paper, Fundamentals of Implementing Real-Time Control Applications in Distributed Computer Systems, outlines a framework for establishing timing requirements of automatic control applications. Further, a real-time behavior model is presented to simplify the analysis of a complex control application prior to implementation. It is further elaborated how the timing requirements are transformed into implementation requirements, and some considerations are made with regard to execution strategies.

The second paper, Design and Programming Tools for Time Critical Applications, gives motivations for specific design tools that can handle the temporal requirements of the application. It describes a graphic-based design and programming environment to assist the development of hard real-time applications. An iterative approach is taken, in which real-time scheduling support is considered from the beginning.

The third paper, Rapid Prototyping of Real-Time Information Processing Units for Mechatronic Systems, is also concerned with development environments and stresses the importance of prototyping tools, especially for the requirements validation phase. The authors present a rapid prototyping environment that supports the designer of application specific embedded controllers and show the use of it in mechatronic designs.

The fourth paper, Implementation of Hard Real-Time Embedded Control Systems, deals with hardware architecture, operating-system and high-level language support to achieve predictability. Special emphasis is devoted to estimation of execution times based on hardware characteristics providing deterministic behavior. The paper has a broad coverage of both hardware and software related issues and hence provides an interesting combination of known concepts towards predictable behavior of embedded control systems.

The fifth paper, HEDRA: Heterogeneous Distributed Real-Time Architecture, provides an application oriented contribution dealing with robot and machine control using distributed and heterogeneous hardware. The focus of the contribution is on openness, flexibility, real-time behavior, minimum interrupt latency and transparency in inter-processor communication. The particular application is a press-brake cooperating with a metal-sheet handling robot.

The sixth paper, Open Embedded Control, shows ways to achieve increased flexibility in mechatronic systems without trading safety, efficiency or simplicity. Instead of only allowing parameter changes, which is the traditional way of offering flexibility, the authors suggest using pieces of executable code as functional operators. Changes can be made even at run-time in the proposed object-oriented framework.

As guest editors it is our hope that this special issue will contribute to the mutual understanding, among real-time system scientists, of the importance of a multidisciplinary approach to the design of real-time control systems. Taking the mechatronics approach we have seen that a close interaction between mechanical engineering, control engineering and computer engineering is a necessary condition for successful real-time control system design. Finally we would like to express our gratitude to the authors for their most interesting contributions and to the Editorial Board for giving us this opportunity.

Real-Time Systems, 14, 219–250 (1998)

Fundamentals of Implementing Real-Time Control Applications in Distributed Computer Systems

MARTIN TÖRNGREN martin@damek.kth.se
Mechatronics lab, Dept. of Machine Design, The Royal Inst. of Technology, S-100 44 Stockholm Sweden

Abstract. Automatic control applications are real-time systems which pose stringent requirements on precisely time-triggered synchronized actions and constant end-to-end delays in feedback loops which often constitute multi-rate systems. Motivated by the apparent gap between computer science and automatic control theory, a set of requirements for real-time implementation of control applications is given. A real-time behavioral model for control applications is then presented and exemplified. Important sources and characteristics of time-variations in distributed computer systems are investigated. This illuminates key execution strategies to ensure the required timing behavior. Implications on design and implementation and directions for further work are discussed.

Keywords: control applications, closed-loop control, multi-rate systems, distributed systems, real-time computer systems, timing analysis, timing jitter

1. Introduction

The successful design and implementation of real-time control applications requires inter-actions between several technical disciplines including mechanical, control and computer engineering. This is especially important in view of current and future mechatronics appli-cations where subsystems are being integrated to increase functionality and reliability and to reduce costs. Integration is done on the control side by inclusion of coordinating control functions and on the computer system side by inclusion of machine embedded networks forming an information backbone in the system.

The aim of this paper is to provide a set of requirements for *real-time implementation* of automatic control applications focusing on discrete-time feedback control systems that employ periodic sampling, also referred to as *sampled data systems*. Such systems are to a very large extent employed in time- and safety-critical parts of mechatronic motion control systems. The requirements are translated into a real-time behavioral model, which provides a framework for design and implementation. Hopefully, the model can provide useful input to research (and evaluation of previous research) on abstract real-time application models, scheduling, synchronization and communication in complex computer systems, which may constitute a combination of multiprocessors and distributed computer systems. These issues are more complicated in the latter category which is of primary concern in this paper. The sources and characteristics of time-variations in distributed computer systems are also investigated. This is done to form a basis for understanding key execution strategies, referring to scheduling, synchronization and triggering policies, required to ensure the required timing behavior of control applications.

The paper is structured as follows. Section 2 surveys related work. Section 3 provides the requirements and the corresponding behavioral model. Section 4 defines important delays

3

and time-variations and investigates their characteristics. Section 5 concludes the paper. The notation used is summarized in the appendix. The behavioral model was first presented in Törngren (1995).

2. State of the Art

2.1. The Multidisciplinary Gap and Research Directions

Four key research directions and their impact on each-other are now introduced.

(1) Computer science and engineering are primarily directed towards the design of general purpose computer systems where high average performance is the leading star, allowing significant time-variations and not providing any real-time predictability. This is a well known fact in the real-time community but still constitutes a problem. Halang (1992) even stated that the use of contemporary computers inhibits real-time control.

(2) Research in the real-time field of computer science is primarily directed towards *discrete-event dynamic systems* where the controlled and controlling systems are modeled in terms of finite state-machines. Most essentially, a real-time control system is perceived as a computer system that has to react to external events within some *stipulated deadline*. A key issue raised and discussed further on in this paper is that the concept of deadlines is necessary but not sufficient for control applications. One attempt to classify control applications with regards to temporal characteristics and requirements is given in Table 1.

(3) Time-delays always deteriorate the performance of a discrete-time control system. Based on points (1) and (2) above it is therefore not so strange that some researchers in control theory have drawn the conclusion that *computer systems are inherently probabilistic in terms of their timing behavior*, see e.g., Belle-Isle (1975), Ray (1994), Shin and Cui (1995), and consequently investigated the influence of stochastic delays on control system stability and performance. Perhaps the earliest research work in this direction was pursued by Belle-Isle (1975) who studied stability for stochastic continuous-time systems. His starting point was the assumption that contemporary computer systems were so unpredictable that a stochastic modeling approach would be suitable! Other researchers have later followed this direction where delays are considered to be stochastic (but usually bounded). Knowledge of their probability distributions is assumed to be available e.g., through statistics gathering. In a properly designed computer control system, time-variations are not by default stochastic in nature on a macroscopic level. It is, nevertheless, important to investigate the sensitivity to and compensation for time-varying delays and data-loss that can occur due to transient failures and bugs. A brief survey of related work is given in section 2.2.

(4) Automatic control theory has, to a very little extent, considered other implementations than dedicated single processor systems. Furthermore, control theory rarely provides advice on how to implement other than single rate control systems. One reason for this is that the analysis and design of multi-rate systems have received relatively little attention compared to single rate systems. The use of multi-rate sampling will most probably increase in the future since it is natural to use and can yield better performance for the same utilization compared to single rate systems (Berg et al., 1988). The theory is however more complicated. Even though text books in digital control theory (see e.g., Åström and Wittenmark

Table 1. Discrete-event vs. discrete-time control systems.

	Discrete-event control systems	*Discrete-time control systems*
Activation	Event-triggered	Time-triggered
Concurrency & dominating communication "paradigm"	Communicating finite state machines & blocking communication.	Communicating periodic threads & non-blocking communication.
Response time requirements	Upper-bound (deadline)**	Upper-bound
Delay requirements in feed-back control	Not (yet) applicable	Constant (upper-bound***)
*Synchronized actions**** *	Specific actions**	Samplings and actuations
Time-variations allowed?	Yes, if less than deadline	Yes, within tolerances

** Requirements on constant delays and synchronization do exist but have rarely been considered explicitly, one exception being in the context of synchronized mode changes (Kopetz and Kim, 1990) and replica determinism as a mechanism to achieve fault-tolerance.
*** The constant delay requirement can under certain conditions be relaxed, see sections 2.2 and 3.
**** Synchronization is in this paper used to refer to actions occurring at the same time with some defined tolerance (*syn' – chron'os*, etymologically traces back to the Greek words "same time").

(1990), Franklin et al. (1990)) emphasize the need for constant feedback delays, a clear picture of real-time behavior requirements has apparently not made its way to the computer science community.

2.2. Control Engineering

As indicated in Table 1, control theory *assumes* a *highly time-deterministic timing* of an implementation. Consequently, very little work has treated *deficiencies in the computer system implementation* of the control system with respect to *time-variations*. Most work has dealt with constant time-delays and continuous time systems. Three types of time-delays can be distinguished (Törngren, 1995):

– *Delays in the controlled system*, for example caused by a system with mass transportation where it is impossible or difficult to obtain a measurement without delay.

– *Sampled data inherent delays* arise due to the periodic operation and the use of a zero order hold in the control system. For small sampling periods this delay can be approximated to $T/2$ (see e.g., Åström and Wittenmark (1990)). Interfaces often required by sampled data systems like analog to digital converters and anti-aliasing filters (often represented as a delay) can be included in this category.

– *Constant computational delays* in the computer.

A time-delay is usually defined as a system which delays a signal but otherwise does not change it. From the theory of continuous-time systems it is well known that a delay system introduces a phase lag, which decreases the stability margin and makes it difficult to design

5

a stable system if a high feedback gain is desired. A control system can be designed as a continuous or discrete-time system. Using ordinary discrete-time control theory, the control system is designed explicitly considering periodic behavior. This facilitates the analysis of systems with *constant time-delays* and makes it relatively straightforward to compensate for the effects on control system dynamics caused by such time-delays (Åström and Wittenmark, 1990; Franklin et al., 1990; Uchida and Shimemura, 1986).

Recently, an increased interest in timing problems with bearing to real-time control systems can be noted (see e.g., Ray and Halevi (1988), Shin and Kim (1992), Wittenmark et al. (1995)). Modeling, analysis and compensation of time-varying discrete-time control systems is difficult and can not readily be done within the framework of ordinary discrete-time control theory. For example, stability properties of a time-varying system can not be directly predicted from the properties of a corresponding time-invariant system (Hirai and Satoh, 1980).

A brief summary of important results is now given.

– Important work has been pursued by Ray and Halevi (1988), and Luck (1989) who investigated the effects of time-varying delays on control system performance and the modeling of time-varying discrete-time control systems. Time-delays and data-loss characteristics of distributed computer systems were investigated but sampling period jitter, response delays and multi-rate systems were not explicitly studied. Execution strategy modifications to reduce or eliminate the time-variations were studied although within a limited scope.

– Time-varying delays have the effect to deteriorate control performance and can potentially cause instability (see e.g., Ray and Halevi (1988), Wittenmark et al. (1995)). The sensitivity of a feedback control system to time-varying delays, or actual constant delays differing from nominal ones, can however vary substantially and among other things depends on the bandwidth of the controlled system and the delay to sampling period ratio. Andreff (1994) presents an example of a simulated motor servo system that performs according to specifications for a nominal constant delay, for which compensation is included, but becomes unstable when the actual delay is small (the difference between actual and nominal delay was within 10% of the sampling period). This emphasizes the fact that also minimum delays are essential for control system implementations and is especially important since computer induced time-variations are normally not considered during control design.

– Time-varying delays and data-loss can be interpreted as computer induced disturbances (see e.g., Törngren (1995), Shin and Cui (1995)). The disturbance interpretation seems particularly valuable for qualitative analysis and intuitive understanding.

– The derivation of *timing tolerances*, stating the allowed deviations from nominal timing and implicitly the sensitivity of a control system towards time-variations, is a new and interesting research area (Wittenmark et al., 1995; Törngren, 1995). Robust control in terms of μ-synthesis and H_∞ design appears to be a promising theoretical approach in this direction. The applicability of the approach is illustrated by Wang et al. (1994) and Wittenmark et al. (1995), where uncertain and/or time-varying delays are modeled as uncertainties in the model of the controlled system. Another related approach is discussed by Shin and Cui (1995) where a performance tolerance in terms of the allowed displacements of poles is used to derive a tolerance on the computational delay.

– Some stability proofs for control systems with time-varying delays exist, with special assumptions regarding execution strategy, delay characteristics and controller structure. For example, Voulgaris (1994) provides a stability proof for a system composed of a sampler and a controller where the controller is triggered to execute by data from the sampler. The sampler to controller delay is assumed to be varying but bounded. The controller uses prediction for delay compensation.

– Minimization of time-variations have been proposed by means of various techniques. For communication between asynchronous periodic tasks, Ray and Halevi (1988) proposed the use of oversampling to eliminate vacant sampling (i.e., the non appearance of a new sample at a controller). If the sampling rate of the sender of data is increased such that $T_{sender} \leq [T_{receiver} + \delta_{min} - \delta_{max}]$, the receiving function always receives a fresh sample during each period, where δ denotes the internal response time of the sending function (see section 4 for formal definition). The approach reduces time-variations at the cost of increased resource utilization.

– Elimination of a time-varying delay by means of synchronization and buffering. Luck (1989) proposed this scheme for two distributed functions, which execute synchronously and where the feedback delay is varying and bounded. The time-varying system is transformed into a time-invariant one by buffering sampled data at the controller such that samples corresponding to the worst case delay are always present at the controller. Provided that other delays are constant, the system is time-invariant and ordinary discrete-time control theory can be employed. To support control applications in time-varying implementations, it is useful if the computer system provides measures of actually experienced delays to the application. With a global clock this is trivial. In other systems this can be achieved by letting the communication system measure significant relative delays occurring during run-time and by including this information in the protocol. By use of actual delay information, actuation instants can then be determined during run-time and constant delays obtained by use of dynamic scheduling (Törngren, 1995).

2.3. *Computer Engineering and Science*

The focus in the real-time computer science community on discrete-event dynamic systems and deadlines can easily be seen in related work and is exemplified in the following.

– Models in structured analysis have been extended towards "real-time systems" by introducing finite state-machines and control-flow in terms of triggers to complement data-flow graphs, see e.g., Hatley and Pirbhai (1987). Timing requirements are limited to response times.

– The application models underlying the MARS system (Kopetz et al., 1989) and HRT-HOOD (Burns et al., 1994) use precedence graphs for real-time behavior specifications.

A response time requirement (deadline) is associated with the complete precedence graph. Precedence graphs are apparently considered to be asynchronous, and while data can be exchanged between precedence graphs constituting a multi-rate system, no timing requirements can be associated with such data-flows.

- The most common interpretation of periodic execution in scheduling theory is an execution that takes place some time during its period. For example, the feasibility test for rate monotonic fixed priority scheduling (Liu and Layland, 1973) guarantees that the execution of independent periodic tasks takes place prior to the end of the period, thus bounding the response time. This work has been extended to cater for more realistic computer activities (see e.g., Audsley et al. (1995) for an overview). The response time focus has however remained. This has the effect to introduce both period and feedback delay variations.

Problems can arise when response time oriented models and associated scheduling theory are used for automatic control applications for which the models are not valid (sufficient). Klein et al. (1994) describe an industrial robot control system which is designed and analyzed using "rate monotonic theory". The servo loops part of the control system are given a deadline equal to their period. The resulting (allowed) timing behavior does not comply with normal timing requirements of sampled data systems. As a simple and obvious example, consider motion control systems where it is common to compute velocity from a position sensor assuming a nominal sampling period. Consequently, the timing requirement appears to be postulated rather than based on control system requirements, compare with sections 2.2 and 3.2.3.

The fact that timing requirements other than response times have been neglected is however being realized (see e.g., Baker and Shaw (1989), Stankovic and Ramamritham (1990), Motus and Rodd (1994)). With relation to this paper, Lawson (1992) discusses the need for synchronization in multi-rate control systems and advocates the use of global static scheduling. Halang (1990) proposed the use of accurately timed computer peripherals combined with a global clock to cater for real-time requirements of control applications. Thus, despite the criticism raised in this paper, many useful concepts with respect to modeling and scheduling have been developed. A brief summary of related work in this very active research field is now given.

2.3.1.   *Models Supporting Constant Delays and Synchronization*

Some models used in static scheduling are based on precedence graphs where the constituting tasks can be equipped with release time and deadline attributes (Xu and Parnas, 1990). Such precedence graphs can be used to specify a constant delay in a single rate control system. The modeling approach is also applicable for priority based scheduling since deadlines shorter than periods and release times have been used in task models.

The Q-model (Motus and Rodd, 1994) allows the specification of a group of synchronous and time-triggered functions, referred to as a synchronous cluster. A time-trigger is defined in terms of a "null channel" and the activation is specified by a *tolerance interval* (defining the maximum activation delay for functions within a cluster with respect to the time-trigger) and a *simultaneity interval* (defining the required precision in the activation of functions within a cluster). The Real-Time Logic used in Modecharts (Jahanian et al., 1988) should also have this ability since it allows the specification of both relative and absolute timing of events.

2.3.2. Scheduling Approaches for Control Applications

Scheduling theory has contributed many useful concepts for timing analysis and design including critical instant analysis, scheduling algorithms, schedule simulation, feasibility tests and response time analysis (see e.g., Audsley et al. (1995) and Stankovic et al. (1995) for an overview). The concepts can be used to guide design and implementation based on both fixed priority and static scheduling. Over recent years it has been understood that scheduling theory, in order to support the design of real-time applications, must be considered in conjunction with policies for triggering, synchronization and communication (see e.g., Kopetz et al. (1989) and Lawson (1992)). The weakest point in the developed theory appears to be the underlying application models.

Attempts to support multi-rate systems include the introduction of so called end-to-end deadlines/delays (Klein et al., 1994; Tindell and Clark, 1994). Klein et al. (1994) take the approach to decompose an end-to-end deadline into local deadlines for periodic functions. This is then treated as a number of local fixed priority scheduling problems. As introduced, the approach is applicable for describing and analyzing bounded end-to-end delay requirements.

Another approach, for example employed in digital signal processing for operations like down- and up-sampling (Lauwereins et al., 1995), is to consider a time-triggered precedence graph which has a period equal to the greatest common divisor (*GCD*) of periods. The tasks part of the graph are specified to execute at an integer multiple of the *GCD*. While this model is simple it is also limited since restrictions on the execution times of functions are introduced (each function must complete within the *GCD*).

3. Modeling Real-Time Behavior of Control Systems

3.1. Control Characteristics in Mechatronics Applications

3.1.1. Basic Modeling Entities: Elementary Functions, Activities and Feedback Loops

In order to model control systems a number of basic modeling entities are first introduced. The primary purpose is to model the real-time behavior of control applications in order to support design and implementation. In a control system, it is assumed that "top-level" functions that provide system services can be decomposed until *elementary functions* have been obtained.

Definition 1. The *elementary function* constitutes a transformation from input to output data. Functions execute when triggered, are connected through unidirectional broadcast

9

channels, and are associated with a simple sequential computational model upon every activation: *read input channels; perform computations*; and *write the computed output to the output channels*. Functions can have local state, i.e., state remaining in between consecutive invocations.

Definition 2. The term *feedback loop* (or path) is used to denote the elementary functions and data-flows between them, that together provide a well defined subset of the control system functionality. This may well include functions executing with different periods, i.e., constitute a multi-rate system. A feedback loop is a control system in which a desired effect is achieved by operating on various inputs to a controlled system until the output, which is a measure of the desired effect, falls within an acceptable range of values (based on the IEEE (1992) definition of a control system).

To facilitate the modeling and specification of multi-rate systems, a further abstraction is introduced.

Definition 3. An *activity* accomplishes *a subset of a feedback loop*, is characterized by one sampling frequency, and is composed of a sequence of elementary functions.

The hierarchy now established is as follows: A control application consists of one or more feedback loops which in the case of a multi-rate system are composed of one or more activities which in turn are built up of elementary functions. For a feedback loop that is implemented within one elementary function, the three actions of the function are interpreted as *sample, compute*, and *actuate*. In the special case of a single rate system, activities constitute feedback loops.

The *feedback loop* and *activity* abstractions are closely related to the specification and modeling of the control application. The *elementary function* abstraction is used in a more general sense in this paper to refer to *control design entities*, i.e., algorithms, samplers and actuators, as well as to *implementation entities*, such as communication subsystem and operating system functions, that are introduced in the implementation stage. In computer science terminology, implementation entities are generated by mapping elementary functions to executable tasks. In a distributed implementation an activity may consequently correspond to one or more tasks.

3.1.2. Characteristics

A number of characteristics of mechatronic motion control systems are illustrated in Figure 1 where the blocks can be interpreted as elementary functions and the dashed ellipses as activities and feedback loops. Different control scenarios range from independent joint control with pre run-time generated references, to complex run-time coordinating control involving inverse dynamics calculations and feedback based synchronization. The scenarios clearly introduce different numbers of functions and interactions between them in the system. Key

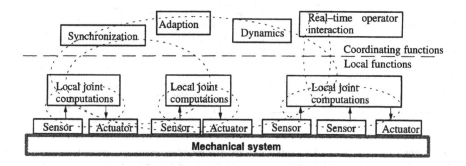

Figure 1. Conceptual structure of and interactions in motion control systems.

characteristics of such motion control systems are as follows:

– Man-machine interfaces often occupy a large percentage of the total software of a
   system and humans are often directly involved in real-time feedback loops, compare a
   pilot or driver.

– Elementary functions, including man-machine interfacing functions, can be divided
   into local and coordinating functions with respect to the actuators and sensors in the
   controlled system.

– Multi-rate sampling is employed. This is natural since subsystems of a controlled system
   typically have different dynamics. In principle all subsystems could be controlled
   with the fastest rate. This, however, may substantially increase the processing load.
   Typically, local functions require higher sampling rates than coordinating ones, which
   in turn require higher rates than for example planning and adaption functions. This
   often leads to a sampling frequency hierarchy. Multi-rate systems may also become a
   practical necessity when a system is built with components from different vendors.

– In a multi-rate system, data-flow takes place between functions (activities) that execute
   with different rates. Motivated by control specific timing requirements associated with
   such interactions the term *multi-rate interactions* is introduced in this paper.

3.2. Sampled Data Systems and Timing Requirements

3.2.1. Basic Operation of Sampled Data Systems

Figure 2 illustrates a single input, single output sampled data system. It is a time-triggered
periodic system where the times when measured signals are obtained are called sampling
instants and the time between successive instants is called the sampling period, T. A
sampling instant is assumed to be given by, $t_k = kT$, where k is an integer. Systems with

11

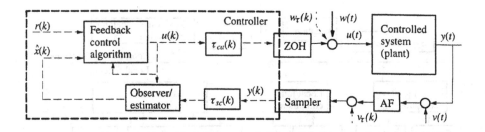

Figure 2. A sampled data system. $\tau_{ca}(k)$ and $\tau_{sc}(k)$ are examples of time-delays introduced by the implementation; $r(k)$ is the discrete-time reference signal; $u(t)$ and $u(k)$ are the continuous- and discrete-time versions of the control signal; $\hat{x}(k)$ is the discrete-time estimated state vector; $y(t)$ and $y(k)$ are the continuous- and discrete-time output signals; $w(t)$ represent modeling inaccuracies and disturbances; and $v(t)$ denotes sensor noise. Time-variations and data-loss can be interpreted as "fictional" disturbances, $w_\tau(k)$, and noise, $v_\tau(k)$ (Törngren, 1995).

non-periodic sampling, for which the theory is not well developed, are not considered in this paper.

A sampled-data feedback system intrinsically constitutes a real-time system. This is manifested by the fact that the *pace of real-time*, e.g., the speed of motion and the rise time of a signal, is used as a basis for determining the required speed of the control system. The choice of sampling period depends on the particular purpose and characteristics of a system. For closed loop control a rule of thumb is to chose 4–10 sampling periods per rise time of the closed loop system, but other factors such as observer dynamics also need consideration (Åström and Wittenmark, 1990).

A sampled data system contains a mix of continuous-time and discrete-time signals, shown with solid and dashed lines respectively in Figure 2. The sampler block obtains outputs from the controlled system and for example represents a sample and hold circuit and an analog to digital converter. AF denotes an anti-aliasing filter, required to eliminate undesirable high frequency contents of $y(t) + v(t)$. The zero order hold block, *ZOH*, includes digital to analog conversion and holds the control signal for one period.

A number of time-delays may exist in the control system as discussed in section 2.2. Of relevance to this paper are delays within the controller that are introduced by the computer system implementation. Such delays have the property that they can vary every sampling period. In Figure 2, a controller to actuator delay, $\tau_{ca}(k)$, and a sensor to controller delay, $\tau_{sc}(k)$, are included. It should be pointed out that the illustrated controller structure is just one example and that the delays may enter in other signal paths depending on how the controller is implemented. The $\tau_{ca}(k)$ and $\tau_{sc}(k)$ delays contribute to the response and feedback delay.

Definition 4. $\tau_c(k)$, the *feedback delay* of a feedback loop, equals the time between related sampling and actuation instants. $\tau_c(k)$ thus equals the effective delay within the feedback path of the controller.

In Figure 3, the execution and timing of a single controller is shown where the feedback

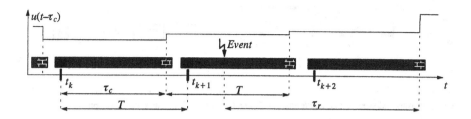

Figure 3. Sampling period (T), feedback delay (τ_c), delayed control signal ($u(t - \tau_c)$) and response delay (τ_r).

delay is constant, i.e., referring to Figure 2, $\tau_c = \tau_{ca} + \tau_{sc}$ is constant. The *response delay*, τ_r, to an event, e.g., a set-point change or a disturbance in the controlled system, is illustrated. A basic assumption in ordinary sampled data theory, that is sufficient for most feedback control purposes, is that of time-invariance. A sampled data system is not in general time-invariant, however, since τ_r depends on the time when the event occurs. In an implementation, the control action can be performed as soon as possible after the corresponding sampling or, alternatively, intentionally delayed and performed e.g., together with the subsequent sampling operation. The former approach is preferable since it reduces response delays.

When the control signal is delayed it is clear that two control signals affect the controlled system during each sampling period, for example $u(k - 1)$ and $u(k)$ during the $k : th$ period in Figure 3, where $\tau_c < T$. With a *state* view of the system it is natural to include the delayed control signal in the model of the controlled system in order to compensate for the delay. With state-feedback, the dynamics of the system can be completely controlled provided that the feedback delay is constant. This in essence means that the destabilizing effect of the time-delay can be eliminated.

3.2.2. Modeling Multi-rate Systems

The theory for modeling and design of multi-rate sampled data systems is based on an assumed deterministic timing, constituting static schedules. The reader is referred to Törngren et al. (1997), Berg et al. (1988), Godbout et al. (1990) for more details on multi-rate control systems.

The sampled data system in Figure 2 can, for example, be turned into a multi-rate system by sampling the observer with a rate $1/T$ and the control algorithm p times slower with $1/(pT)$, where p is a positive integer. This has the effect to introduce a "fast-to-slow" sampling rate transition for the data-flow. Another example of such a multi-rate system is where a digital filter precedes a feedback algorithm and in conjunction with an analog filter provides anti-aliasing. Many other types of transitions, e.g., "fast-slow-fast" transitions in a cascade coupled systems, are also possible.

A multi-rate system is fundamentally characterized by the relation between periods. Given a set of elementary functions with periods $\{T_1, \ldots, T_n\}$ in increasing order, the ratio between periods is of interest. A special but interesting case is when the periods are related as

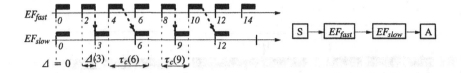

Figure 4. Phasing (Δ, skew ($\Delta(j)$)) and feedback delay ($\tau_c(j)$) for the multi-rate system example.

follows: $T_{i+1} = N \cdot T_i$, where $GCD(T_1, \ldots, T_n) = T_1$, $LCM(T_1, \ldots, T_n) = T_n$, and N is an integer, e.g. $N = 2$ yields periods $\{T_1, 2T_1, 4T_1, \ldots\}$. Such period relations not only simplifies modeling and design of the multi-rate control system but also the implementation. Practical implementations will clearly aim for such period relationships. Such relations may, however, not always be possible or preferable to choose from a feedback control point of view. A more general case is as follows. $T_0 = GCD(T_1, \ldots, T_n)$ and $T_1 = N_1 \cdot T_0$, $T_2 = N_2 \cdot T_0, \ldots, T_n = N_n \cdot T_0$, where the N_i's are integers. For example, $\{2T_0, 5T_0, 9T_0\}$ and $LCM(2T_0, 5T_0, 9T_0) = 90T_0$. It follows that the ratio between successive periods in this case is a rational number.

A further fundamental characteristic is that of synchronism.

Definition 5. Activities are *synchronous*, also referred to as *synchronous execution*, when "related" period start points of the activities always are separated by a known constant called *phase*, Δ, within a specified tolerance. For synchronous execution in multi-rate systems, "related" periods is interpreted such that the phase refers to the instants of the "major period", i.e., the *LCM* of periods. The definition is equally valid for elementary functions.

Consequently, in *asynchronous* execution the period start points of activities do not have a guaranteed relation. The term *skew* is used to refer to a time-varying phase between period start points.

Definition 6. The *skew*, $\Delta(j)$, between two communicating periodic elementary functions, EF_1 with period T_1, and EF_2 with period T_2, where EF_1 is communicating data to EF_2, is defined based on the sampling instants of EF_2, the receiving function. The skew for each instant of EF_2 equals the interval between this instant and the closest previous sampling instant of EF_1. The definition is equally valid for activities.

The definition of skew is valid for both single and multiple clock systems. For asynchronous functions with the same period the skew is time-varying due to clock drift. Unless period ratios are integer constants, the phase is also time-varying in synchronous multi-rate systems. As an example, consider two synchronous activities, EF_{fast} and EF_{slow}, with periods $\{T_{fast} = 2, T_{slow} = 3\}$, and zero phase, where the loop is closed over the fast-to-slow rate transition, see Figure 4. The skew becomes $\Delta(j) = 0$, $\Delta(j+1) = 1$, repeating over and over.

In order to treat the timing behavior of multi-rate systems it is also necessary to distinguish between the *feedback delay*, Definition 4, and the *activity duration*.

14

Definition 7. The *activity duration* equals the time duration of an activity, counted from the starting trigger of the activity to the completion of its final constituting elementary function.

The feedback delay and the activity duration are equivalent only for single rate systems. In case of a multi-rate feedback-loop system, the loop is composed of one or more activities. Consider again the example of Figure 4 where both activities have a duration of one time unit. It follows that $\tau_c(j) = 3$, $\tau_c(j + 1) = 2$, repeating over and over.

The timing requirements of a multi-rate control system can consequently be divided into requirements on *synchronous execution* and *constant activity durations*. The "*sampling schedule*" repeats with the *LCM* of the periods and the *GCD* is the interval over which all control signals and outputs from samplers are constant. The system can be analyzed at *LCM* and/or *GCD* instants.

3.2.3.   Timing Assumptions: Real-Time Behavior Requirements

The timing assumptions made in discrete-time control theory introduce the following requirements:

Constant sampling period. Sampling is performed at equidistant time instants given by the sampling period.

Activity synchronization. There is a common clock that controls the execution of different components in a synchronized manner. The different activities in a multi-rate system are perfectly synchronized with a constant phase relation at *LCM* instants.

Constant activity duration. The delay between related sampling and actuation instants for an activity is constant, i.e., actuation is performed at equidistant time instants given by the sampling period. The requirement is valid for both single and multi-rate systems.

For single rate and synchronous multi-rate systems, where the period ratios are integers, the third requirement can be translated into a requirement on **constant feedback delays**.

For multi-rate systems, where the ratio between successive periods is a rational number, **a set of deterministic constant feedback delays** are obtained in the system that repeat over and over. For an activity, i, that interacts with faster activities in a feedback loop, LCM/T_i feedback delays are obtained. The delays depend on the phasing, the periods and the duration of the activities.

Deviations from nominal timing are *only* allowed as long as they do not interfere with the given requirements. For example, for a single rate control system with a nominal sampling period and feedback delay, this implies that time-variations can only be allowed if they are sufficiently small considering the robustness of the control system towards time-variations.

3.2.4. Implementation Goals and Constraints for Periods, Delays and Tolerances

For real-time implementation, analysis and specification of sampling periods, constant delays of activities and loops, and suitable optimization criteria are required. To facilitate the implementation of the deterministic timing requirements of sampled data systems, the concept of tolerances, further detailed in the following section, should be very useful.

Sampling periods are derived on a basis of the characteristics of the controlled system and the control objectives. Normally, a range of acceptable sampling periods can be identified which satisfy system requirements, i.e., $T_{short} \leq T \leq T_{max}$. Based on the execution time, C, of an elementary function and its allowed utilization, U_{allow}, a necessary lower bound is given by, $C/U_{allow} \leq T_{min}$. The range of available sampling periods can be used to facilitate implementation while at the same time meeting control requirements. This clearly illustrates the need for interactions between control and computer engineers.

Timing tolerances capture the fact that a limited deviation from nominal timing referring to equidistant sampling periods, constant feedback delays and synchronization can be allowed. The derivation of tolerances is a topic for further research, in particular with relation to multi-rate systems and different control design approaches (Törngren et al., 1997). Tolerances provide implementation flexibility by allowing alternatives to static scheduling and synchronous execution. They may also be useful to optimize static schedules. Since tolerances provide information about timing criticality they should be useful in exception handling and in graceful degradation.

Feedback delays are always desirable to minimize, thus providing an optimization problem for the implementation. In certain cases a control system may already have been designed for a particular constant delay. In other cases, delays may not have been considered at all. This again illustrates the need for interactions between control and computer engineers; in this case actually obtained delays should be fed back to control design engineers so that the effects of the time-delay can be investigated and compensation properly included in the discrete-time control system. Furthermore, allocation and scheduling optimization can be carried out to reduce feedback delays.

Specification approaches for multi-rate systems are discussed in the following section. For the case of rational period relations, specifications could be given in terms of activity durations *or* in terms of a set of control delays.

3.3. A Real-Time Behavior Model of Control Systems

A timing behavior model that captures the above presented requirements of real-time control systems is outlined. The most essential ingredients are *precisely time-triggered actions* with *tolerances* and the specification of *multi-rate interactions*. The main purpose of the model is to facilitate the description of application timing requirements and to support the multidisciplinary design and implementation process.

3.3.1. Precisely Time-Triggered Actions and Tolerances

The notion of precisely time-triggered actions can be introduced for both relative and absolute timing constraints. The term "precisely time-triggered" is used since it corresponds to the assumption in sampled data theory that sampling and actuation actions occur exactly at specified time instants. Translating this into the non-exact reality it is natural to introduce *tolerances*, i.e., $t_{nominal} \pm tolerance$, where $t_{nominal}$ refers to a specified absolute time. For sampled data applications a relative accuracy (bounded clock drift) is sufficient. Consequently, *precisely time-triggered* sampling and actuation actions can be defined with *tolerances* as follows

$$\text{Precisely time-triggered sampling actions: } t_{sample}(k+1) = t_{sample}(k) + T \pm tol_T \quad (1)$$

$$\text{Precisely time-triggered control action: } t_{actuate}(k) = t_{sample}(k) + \tau_c \pm tol_\tau \quad (2)$$

Eq. (1) and (2) describe timing requirements which must be met by an implementation. That is, actual periods and control delays are allowed to vary within their respective tolerances. Consequently, Eq. (1) and (2) describe the required relative timing between successive sampling instants, and related sampling and control instants respectively, i.e., for example:
$$T - tol_T \leq t_{sample}(k+1) - t_{sample}(k) \leq T + tol_T$$

In addition to Eq. (1) and (2), a synchronization tolerance is normally required. Törngren (1995) used period and delay tolerances to define synchronization requirements for precisely time-triggered actions in a multivariable control system (i.e., a system that has several inputs and outputs, therefore requiring synchronization of sampling and actuation actions).

Recent research reveals that the derivation of synchronization tolerances and their relation to sampling period and delay tolerances is rather complicated in the general case (Törngren et al., 1997). Some of the problems are briefly illustrated in the following.

For a multivariable single rate control system we may derive tolerances on the feedback delay and the sampling period assuming synchronous execution. If we would like to consider an asynchronous distributed implementation of the control system, we also need to investigate a synchronization tolerance. Whereas theoretical approaches to investigate sampling period and feedback delay tolerances are emerging there is currently no other available approach then simulation and tests to derive synchronization tolerances.

Consider a robot arm composed of two dynamically coupled links where multi-rate control is used since the dynamics of the two degrees of freedom differ. To compensate for the "disturbance" forces caused by the link coupling, some data such as actual positions and applied control signals would need to be exchanged between the feedback loops. An actual delay differing from the nominal feedback delay of one loop will clearly affect the other loop, and may do so not only through the mechanical coupling but also by potentially violating the assumed internal timing behavior in the multi-rate interaction.

A pragmatic approach to overcome these difficulties is clearly to ensure synchronous execution such that the synchronization tolerance is a fraction of the corresponding period or delay tolerance. The tol_τ and tol_T tolerances can then be derived assuming synchronous execution.

The synchronization tolerance, tol_{SYNC}, and its relation to tol_T is given by Eq. (3).

$$\varepsilon_s(j) \leq J_{max,s} + \xi \leq tol_{SYNC} \ll tol_T \quad (3)$$

17

For the implementation of synchronized sampling actions this means that the difference between two related sampling instants, $\varepsilon_s(j)$, is bounded by ξ, the precision of the time-trigger(s) providing synchronization, and, $J_{max,s}$, the maximum jitter that can be encountered (further elaborated in section 4).

For cascaded multi-rate systems, the feedback delay of a slower (cascaded) loop can be used to investigate whether synchronous execution is required or not, as will be exemplified in section 3.3.3.

3.3.2. Specifying Triggers for Elementary Functions and Control Flow for Activities

Activation triggers for elementary functions are now introduced. A trigger is characterized by the following attributes: {*Type, Source,* [<] *Period:* [[<]τ], [*tol*]}, where [ ] denotes optional and < is an optimization directive. Two *types* of triggers exist: event-triggers (*ET*) and time-trigger (*TT*). *ET*'s are not further considered in this paper. A trigger *source* is specified as a *clock* with a particular name. The same trigger source can be used for several functions if synchronous execution is desired.

Within one activity all time-triggers must have the same period and clock, but different constant delay specifications, [[<]τ], are allowed. This is primarily used to specify constant feedback delays and/or activity durations. To avoid ambiguity, if it is not clear from the context, a subscript d can be used to refer to an activity duration, $\tau_{d,i}$, whereas $\tau_{c,i}$ as usual refers to a feedback loop delay. For multi-rate systems it is also desirable to be able to describe a set of control delays; the required syntax is not elaborated in this paper.

Typically, a control activity will include two precisely time-triggered actions (sampling and actuation) in order to specify a constant feedback delay and/or duration. Functions part of activities are thus related through precedence relations, specifying execution order, and time-triggers, specifying precise timing.

When the delay specification is used in a single rate system or for an activity duration in a multi-rate system, a pair of time-triggers are used to define a time-window during which intermediate functions and data-flows must be carried out.

In case the delay specification refers to a feedback delay, the specification naturally refers to the sampling and actuation actions of the corresponding feedback loop which may include more than one activity. An example is given in the following section.

The delay specification could conceptually also be used to specify phase relations for synchronous functions not part of the same activity, e.g., useful in implementation refinements. The common requirement of a constant but minimized feedback delay is specified by [< τ]. The tolerance specification, [*tol*] is interpreted depending on its position within an activity. The tolerance refers to the period when given for the first function in an activity, otherwise to the delay. The tolerance is interpreted as a synchronization requirement when the same trigger source is used for several functions.

A precedence relation between elementary functions A and B corresponds to B being event-triggered by the completion of A. For control applications it is common that control- and data-flow coincide. However, the separation of control- and data-flow specifications is still preferred for clarity and to be able to appropriately model interactions in multi-rate systems.

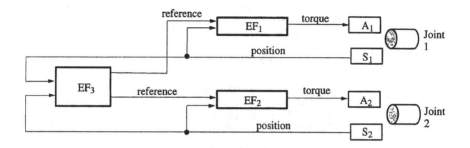

Figure 5. Data-flow graph of the example multi-rate control system.

3.3.3. Control-Flow and Timing Specifications for Multi-rate Systems

Törngren (1995) identified three alternatives for describing the real-time behavior of multi-rate control systems. The first obvious approach is to model the timing in one precedence graph extended to the *LCM*. That is, all activities part of a multi-rate system are modeled together—*merged*—to form one comprehensive activity which is periodic with *LCM*. This approach is similar to specifying a static schedule where all release times and activity durations during the *LCM* must be included.

The two other approaches rely on *activity* and *feedback loop* modeling respectively. These approaches provide more concise specifications of multi-rate systems. Also, they can be used to specify both synchronous and asynchronous multi-rate systems, as well as traditional computer science oriented models based on precedence graphs. Since *activity* and *feedback loop* modeling provide better support for design and implementation, the rest of this section focuses on these approaches. The example multi-rate control system in Figure 5 is used to illustrate the approaches. The example system involves two "local" elementary functions, EF_1 and EF_2, corresponding to two motor servos, and one coordinating function, EF_3, that performs feedback based coordination. It is assumed that EF_1 and EF_2 have period T and EF_3 period T_3, where $T_3 > T$. $S_{i:1..2}$ (similarly for $A_{i:1..2}$ and $EF_{i:1..2}$) is used to denote S_1 and S_2. Three feedback loops can be identified in the system, two "local" and one coordinating. Each local loop obtains samples from its sensor, S_i, performs the local function, EF_i and then actuation, A_i. The coordinating loop stretches from the sensors to EF_3, and further on through the reference path connection, to local functions and actuation. The system can consequently be modeled as three activities (or alternatively as two if the local functions are modeled together).

In Figures 6 and 7, control-flow and timing specifications are illustrated for the example system. Circles denote elementary functions and the notation introduced in 3.3.2. is used to specify time-triggers (*TT*), trigger sources (*CLK*) periods, delays (τ) with minimization directives (<), and tolerances (*tol*). In Figure 6a, the same feedback delay with minimization directive and tolerance are specified for the local activities which are also specified to execute synchronously. Alternatively, if different delays are required, the controllers are specified separately. Horizontal arrows are used to denote precedence relations and vertical

19

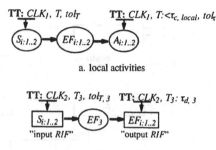

a. local activities

b. coordinating activity

Figure 6. Timing and control flow specification of the example system using rate interfacing functions (RIFs).

arrows time-triggers. For example, for the local activities in Figure 6a, the precedence relation together with time-triggers specify that the local functions and data-flows should be carried out between the given time-triggers. The intermediate actions can consequently be associated with an overall deadline equaling $\tau_{c,local}$.

3.4. Activity Modeling and Rate Interfacing Functions

The basic idea in this approach is to model activities *separately* and to introduce *rate interfacing functions* (*RIFs*), which together with the timing specifications for activities ascertain that the delays in data-flows between activities are sufficiently bounded or constant. The approach is used in Figure 6b for the coordinating activity. Boxes in control-flow graphs are used to denote *RIFs*. A *RIF* retrieves or provides data from or to an elementary function that belongs to another activity, subject to the given timing specifications. In the example the delay specification for the output *RIF* in Figure 6b refers to the duration, $\tau_{d,i}$, of the coordinating activity. The input *RIF* in Figure 6b is interpreted as the retrieval of the output data available from functions S_1 and S_2 at times (kT_3) according to CLK_2, where the data then available at $S_{i:1,2}$ is given by the specification of the local activity. The output *RIF* is interpreted as the provision of the output data from EF_3 to functions EF_1 and EF_2 such that the data is available to the receiving functions for use at times $(kT_3 + \tau_{d,3})$ according to CLK_2. In the implementation, the precedence graphs need to be refined and the *RIF*'s appropriately replaced by elementary functions that perform remote and/or local read and write operations depending on if the activities are implemented in the same or different nodes of a distributed system.

When $CLK_1 \equiv CLK_2$ in Figure 6, the coordinating vs. local activities are defined to be synchronous. The combination of constant activity durations and synchronous execution, perfectly matches the modeling requirements of multi-rate systems. In the case where T_3/T_1 is a rational number, the feedback delay of the coordinating loop is time-varying but the schedule is deterministic and the same delays repeat each *LCM*. As mentioned earlier, the specification of the timing attributes of such a multi-rate system could be done in two ways; either by specifying the duration for the coordinating activity as in Figure 6b, or by specifying the set of feedback loop delays. As indicated in the example in section 3.2.2. durations and feedback delays can be translated into each other.

Note that the suitability for using an activity duration vs. a feedback delay in the general case would depend on the type of multi-rate system configuration and also on execution strategy considerations. The specification of activity durations is in principle sufficient when a deterministic implementation is at hand (e.g. corresponding to static scheduling). It is still advisable to specify the feedback delay and its tolerance explicitly since this on one hand provides more degrees of freedom for the implementation, and on the other hand requires derivation of timing tolerances, providing important information of the system sensitivity.

If the coordinating vs. local activities are defined to be asynchronous, i.e., meaning that they use different clocks, each data item that is retrieved through the input RIF of Figure 6 may in principle have been created up to one sampling period, $T + tol_T$, ago. There is a detection delay of up to $T + tol_T$, associated with the delivery through the output RIF. Consequently, even if the activity durations are constant, the feedback delay of the coordinating loop will be time-varying due to asynchronous execution. For the specification given in Figure 6b the upper bound is given by $(2T + 2tol_T + \tau_{d,3})$.

For an asynchronous solution, however, the designer should specify the required feedback delay(s) and tolerance(s) of the coordinating loop. For example, when T_3/T_1 is an integer, the specification for the second trigger in Figure 6b should instead read: CLK_2, T_3, $t_{c,3}$, $tol_{\tau,3}$. The timing tolerance, $tol_{\tau,3}$, can then be compared against the expected variation of the feedback delay due to asynchronous execution, to determine whether asynchronous execution is feasible or not.

Rate interfacing functions are straightforward to introduce for other multi-rate interactions. For example, assuming a fast-to-slow transition from the *observer* to the *control algorithm* function in Figure 2, it would be sufficient to introduce one RIF, which could be placed in either of the two activities. The timing of the data-flow from the *control algorithm* to the *observer* could also be specified using a RIF. Normally, however, this would be defined from the activity or loop specifications.

3.5. Feedback Loop Modeling

In the interesting practical case when the period ratios are integers, an alternative where RIFs are not required in the specification is available. The idea is to model feedback loops rather than activities. This is facilitated since feedback delays are constant in this special multi-rate case. Figure 7 illustrates the use of feedback loop modeling for the coordinating loop of the example system. The modeling of the local loops is identical to that given in Figure 6a.

Feedback loop modeling means that a complete path, which may involve more than one activity, is modeled and that the activities are synchronous. In the perspective of activities this means that a sampling frequency hierarchy is adopted and that an activity at a particular "level" is specified by including all related faster activities part of the loop. As a consequence, an elementary function can be used in several feedback loop specifications, implying that the specifications are *overlapping*. For the example system it is evident that

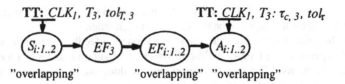

Figure 7. Timing and control flow specification of the coordinating loop of the example system using *feedback loop modeling.*

this is the case for $S_{i:1..2}$, $EF_{i:1..2}$ and $A_{i:1..2}$. A function can because of this be associated with several sets of timing requirements which must be resolved during scheduling and optimization. As mentioned earlier, a feedback loop model can be converted into an activity model.

4. Characteristics of Time-Variations in Distributed Computer Control Systems

4.1. Sources of Time-Variations and Execution Strategies

The timing behavior of an implementation in general depends on a number of factors including:

- *The structuring and allocation of the control application.* A natural primary approach is to use locality based distribution such that feedback loops are executed at relevant I/O nodes. Nevertheless, a loop may need to be closed over a network. This degree of freedom is further elaborated elsewhere (see e.g., (Törngren and Wikander, 1996)).

- *Execution strategy considerations.* In a distributed system, suitable choices of policies with respect to triggering, synchronization and scheduling (for execution and communication) are essential for the timing behavior of the system. Fault handling policies are equally important but outside the scope of the paper. A set of selected policies is referred to as an *execution strategy*. There is a range of execution strategies that can provide predictable real-time behavior.

- *The predictability of the hardware and software "components".* Varying execution and communication time $C_{min} \le C(k) \le C_{max}$, and clock drift fall into this category. The sources of time-variations due to component characteristics are not considered further in this paper.

The latter two constitute sources of time-variations during real-time operation, assuming that dynamic allocation is not used. In the following investigation the focus is on time-variations due to execution strategy choices. Consider the execution of a number of elementary functions that form a control system. Fundamental execution scenarios are as follows:

- The functions are either time-triggered or event-triggered.

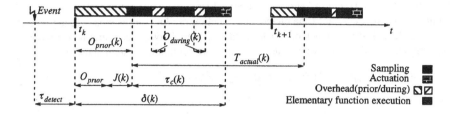

Figure 8. Internal response time ($\delta(k)$), jitter ($J(k)$) and overhead ($O_{prior}(k)$ and $O_{during}(k)$) for a single elementary function.

– The time-triggering is based on a local or a global time base.

– The event-triggered function is either truly aperiodic or sporadic. The former type is always enabled for execution once a relevant event occurs. The latter is only enabled provided that a specified minimum period has expired since the last event occurred (Mok, 1983).

– The system is completely statically scheduled, completely scheduled based on static priorities, or scheduled using some intermediate "hybrid" policy.

4.2. Delays and Time-Variations in Control Systems

4.2.1. Definitions for a Single Elementary Function

The *internal response time*, the *overhead*, and *sampling period variations*, termed *jitter*, are introduced and illustrated in Figure 8. For simplicity, but closely corresponding to real implementations, it is in the following assumed that sampling and actuation actions take negligible time to perform, that elementary functions conform to Definition 1, that resource utilization is suitably lower than 100% and that "messages" between elementary functions are delivered in order. The presented equations can be extended to cater also for other cases.

Definition 8.   The *internal response time*, $\delta(k)$, of an elementary function is counted from the point in time of the starting trigger until the function has completed execution. Predictability implies that $\forall j: \delta_{min} \leq \delta(j) \leq \delta_{max}$. For a particular system, $\delta_{max}/\delta_{min}$ is a measure of the time-variations that can be expected due to application characteristics, the execution strategy and component predictability.

In scheduling theory, the response time of a task to an event is computed as $\delta = C + B + I$, i.e., the sum of the execution time of the function; potential blocking, accounting for time when a lower priority task suspends a higher priority one; and interference, accounting for the execution of activities with higher priorities. For an elementary function, C can denote the duration of execution and/or communication.

23

Definition 9. The *response delay*, τ_r, is counted from the point in time when an external event of interest occurs until the elementary function produces a response (actuation). A detection delay is due to a lack of synchronization, either between cooperating activities in the computer system or between external events and a computer system activity.

$$\tau_r(k) = \tau_{detect}(k) + \delta(k), \text{ where } 0 \le \tau_{detect}(k) < T$$

To express the feedback delay and sampling period jitter, independent of scheduling approach, the system and application overhead need to be considered.

System overhead refers to blocking caused by the operating system, e.g., preemption, dispatching and interference caused by operating system tasks such as the clock tick task.

Application overhead refers to the interference and blocking caused by the multiplexing of several activities on one processor. While these overheads normally are associated with "dynamic" (run-time) scheduling, they can also be interpreted for static scheduling as statically scheduled blocking and interference. Regardless of the sources of blocking and interference, overhead is grouped into $O_{prior}(k)$ and $O_{during}(k)$.

Definition 10. $O_{prior}(k)$ represents the overhead that delays the start of the elementary function, i.e., it takes place *prior* to sampling. $O_{during}(k)$ represents the overhead taking place *during* the execution of the function. Each overhead can be divided into a static and a varying part.

The feedback delay for a single elementary function according to Definition 4 can be expressed as:

$$\tau_c(k) = O_{during}(k) + C(k) \tag{4}$$

Definition 11. *Jitter*, $0 \le J(k) \le J_{max}$ refers to non-intentional variations in the sampling period. Jitter constitutes the time-varying part of $O_{prior}(k)$, i.e. $O_{prior}(k) = O_{prior} + J(k)$, where $O_{prior} \ge 0$ is a constant.

If clock drift is neglected it follows that the actual sampling and actuation instants, $t_{sample}(k)$ and $t_{actuate}(k)$, respectively, can be expressed as follows:

$$t_{sample}(k) = t_k + O_{prior}(k) = t_k + O_{prior} + J(k) \tag{5}$$
$$t_{actuate}(k) = t_k + O_{prior}(k) + \tau_c(k) = t_k + O_{prior} + J(k) + O_{during}(k) + C(k) \tag{6}$$

IEEE (1992) defines jitter as follows: "*Time-related, abrupt, spurious variations in the duration of any specified related interval*". The intervals could relate to the sampling period and to a specified delay. In this paper, however, *jitter* is used to refer to *sampling period variations*. There are two main reasons for this. In control theory, constant and varying feedback delays are well established terms whereas spurious variations of sampling periods have not been studied. Moreover, the two types of time-variations (jitter vs. a time-varying delay) affect the feedback loop performance in different ways (Törngren, 1995). This can easily be seen by investigating how T and τ_c affect the coefficients of a discrete-time closed

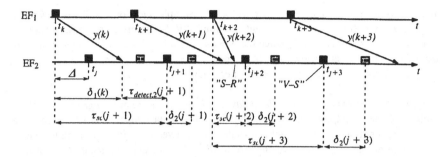

Figure 9. Timing relations between two periodic elementary functions.

loop system. A further indication is given by the following consideration. The case with $O_{prior} > 0$ and $J(k) = 0$ corresponds to a sampling system that has a static offset. This system is thus time-invariant and poses no direct timing problems. When the time-varying part in Eq. (6) is zero, however, a constant feedback delay remains which does affect the dynamics of the controlled system and needs to be compensated for unless the delay is negligible.

4.2.2. Definitions for a System with Two Elementary Functions

The basis for delays in complex systems is the interaction between two elementary functions. Figure 9 depicts such a system. The first elementary function, EF_1, performs sampling and then delivers the sampled data to the second function, EF_2, which reads its input data at EF_2 sampling instants, performs a computation and then actuation. For the purpose of the following definitions it is assumed that both EF_1 and EF_2 are periodic with a period, T. Note that the internal response delay of EF_1 includes the actual delivery of data to EF_2, compare with $\delta_1(k)$ in Figure 9.

There are two main cases, synchronous and asynchronous execution. To formalize the definition of skew given by Definition 6, $t_{k^*(j)}$ is introduced to denote the sampling instant of EF_1 that is closest to the following sampling instant j of EF_2. The skew can then be expressed as:

$$0 \leq \Delta(j) = t_j - t_{k^*(j)} < T, \text{ where } k^*(j) = max\{k: t_k \leq t_j\}, \tag{7}$$

In Eq. (4) the feedback delay is defined for one elementary function. For the case where a feedback loop is divided into two periodic elementary functions, the relevant interval from sampling to actuation is more difficult to establish. The reason for this is that *sample rejection* and *vacant sampling* can occur (the terms were introduced by Ray and Halevi (1988)). Sample rejection refers to the case when more than one data item arrives to EF_2 during a period and the particular case when a later value overwrites the previously received one. Sample rejection according to this definition is thus equivalent to the case when a data item is lost, and there could be other reasons for this. Sample rejection, "S-R", is illustrated in Figure 9 prior to instant t_{j+2}. Vacant sampling, "V-S", refers to the case where no new data-item has arrived to EF_2 since the previous sampling instant. Vacant sampling is illustrated in Figure 9 at instant t_{j+3}. To define the feedback delay for such a system it is convenient to introduce the sensor to controller delay.

Definition 12. The *sensor to controller delay*, $\tau_{sc}(j)$, is a measure of the *age* of the freshest sample available at sampling instant j, accounting for the data delay in interactions between two periodic functions. $\tau_{sc}(j)$ is defined as the time between the j:th sampling instant of EF_2 when a sample $y(k''(j))$ is obtained from EF_1, and the sampling instant $t_{k''(j)}$ of EF_1 when $y(k''(j))$ was sampled by EF_1.

The sensor to controller delay can be expressed by:

$$\tau_{sc}(j) = \delta_1(k''(j)) - O_{prior,1}(k''(j)) + \tau_{detect,2}(j)$$
$$= p_{min}(j)T - O_{prior,1}(k''(j)) + \Delta(j) \tag{8}$$

where $k''(j) = k^*(j) - p_{min}(j)$, and $p_{min}(j) \geq 0$, is derived as follows:

$$p_{min}(j) = min\{p \geq 0, \delta_1(k^*(j) - p) \leq pT + \Delta(j)\} \tag{9}$$

Example: From Figure 9 it is clear that: $\delta_1(k) < [\Delta+T]$ and $\delta_1(k+1) > \Delta \rightarrow \tau_{sc}(j+1) = [\Delta + T]$. For the $(j + 1)$:th instant of EF_2, $k^*(j + 1) = k + 1$, and $p_{min}(j + 1) = 1$.

The feedback delay at a particular sampling instant of EF_2 depends on the "age" of the data then available at EF_2, described by the sensor to controller delay. See Figure 9 for examples.

$$\tau_c(j) = \tau_{sc}(j) + \delta_2(j) \tag{10}$$

4.3. Characteristics of Time-Variations

The characteristics of time-variations are discussed in the following for a feedback loop system composed of two elementary functions, EF_1 and EF_2, with sampling periods T_1 and T_2 respectively.

For simple systems, the characteristics of time-variations are often straightforward to analyze a priori. Consider for example priority based scheduling, a set of periodically activated control tasks which each implement a feedback loop, one sporadic interrupt routine and a clock tick task. Typically, a number of discrete values for jitter and the feedback delays can be determined (corresponding to whether an interrupt occurred or not, whether the clock tick occurred or not, etc.).

4.3.1. Sampling Period Variations—Jitter

Unless a precisely time-triggered action is implemented properly, sampling periods will vary due to jitter. From Eq. (5) it follows that the actual period may take on values in the range $T \pm J_{max}$. Furthermore, several consecutive periods can be less or larger than T.

For a single elementary function, a specified period tolerance can be translated into a jitter requirement, compare with Eq. (1). This is possible since clock drift in most practical cases is negligible during a period. Compare with $T_{actual} = T(1 \pm \varrho)$, where ϱ is the maximum clock drift in [s/s]. Often $\varrho T \ll T$, e.g. for quartz clocks with ϱ in the order of 10^{-5} s/s.

Jitter should be eliminated by proper design and this is conceptually straightforward, either based on a direct hardware implementation, or by use of suitable scheduling polices. If this is the case, a tolerance specification for synchronized precisely time-triggered actions can be translated into a requirement on the precision of the synchronization mechanism, compare with Eq. (3).

4.3.2. Response Delay Variations

The response delay to an event detected by EF_1 can be expressed as

$$\tau_r = \tau_{detect,1}(k+1) + \delta_1(k+1) + \tau_{detect,2}(j_{min}) + \delta_2(j_{min}) \tag{11}$$

In Eq. (11), j_{min} refers to the instant when the sample corresponding to the event is sampled by EF_2. It is clear that to reduce the response delay it is necessary to use a sufficiently small sampling period. The response delay depends on the internal response time of EF_1 and the skews encountered with respect to EF_2 (causing the second detection delay). For the response to a particular event it is clear that the best case is when the sample corresponding to the event arrives at EF_2 precisely at a sampling instant of EF_2. The bounds on the response time delay are given by

$$\forall k : \delta_{min,1} + \delta_{min,2} \le \tau_r(k) < \delta_{max,1} + \delta_{max,2} + T_1 + T_2 \tag{12}$$

In Eq. (12), T_1 and T_2 are due to the worst case of the detection delays. It can be concluded that relatively large variations in response times can be obtained, relative to the sampling periods. The effect of making EF_2 aperiodic is to eliminate the detection delay, i.e., $\tau_{detect,2} = 0$ in Eq. (11). Consequently the worst case term T_2 is eliminated from Eq. (12).

4.3.3. Feedback Delay Variations

First, consider a single elementary function. The bounds of the feedback delay depend on how actuation is performed. If actuation is performed as soon as possible, without a separate time-trigger, the bounds are established from Eq. (4). If actuation is performed together with sampling, actual actuation instants are given by

$$t_{actuate}(k) = t_{k+1} + O_{prior}(k+1) = t_k + T + O_{prior} + J(k+1) \tag{13}$$

The nominal feedback delay then equals the sampling period. In this case the delay is affected by the jitter, i.e., $\tau_c(j) = T \pm J_{max}$. If actuation is separately time-triggered it is appropriate to model actuation as a separate elementary function (compare sections 3.3.2 and 3.3.3).

For a system of two periodic elementary functions it follows from Eq. (8) and (10) that the bounds of the feedback delay are given by

$$\forall j: \delta_{min,1} - O_{prior,1} + \delta_{min,2} \leq \tau_c(j) < \delta_{max,1} + \delta_{max,2} + T_1 - O_{prior,1} \qquad (14)$$

The best case occurs when $[O_{during,1}(k) + C_1(k)]$ is minimal. It should be noted that this may well occur while $O_{prior}(k)$ takes on its maximum value. The worst case is derived for asynchronous functions and occurs under the following condition: the skew is such that a sample from instant i of EF_1 delayed by $\delta_{max,1}$ precisely misses a sampling instant j of EF_2. Since messages are assumed to be delivered in order it follows that subsequent samples also will miss instant j. This means that the sample from instant $(i-1)$ of EF_1 will be sampled by EF_2 at j, thus yielding the two final terms in Eq. (14).

When the skew is known at sampling instant j of EF_2, the worst case upper bound is as follows:

$$max[\tau_c(j)] = \left\lceil \frac{\delta_{max,1} - \Delta(j)}{T_1} \right\rceil T_1 - O_{prior,1} + \Delta(j) + \delta_{max,2} \qquad (15)$$

Eq. (15) follows directly from Eq. (8) and (10) where $p_{min}(j)$ is derived from Eq. (9) by replacing $\delta(k^*(j) - p)$ with $\delta_{max,1}$. I.e.: $p_{min}(j) = min\{p \geq 0, [\delta_{max,1} - \Delta(j)]/T_1 < p\}$. The ceiling brackets follow since $p_{min}(j)$ is an integer. Uncertainty due to clock drift can be accounted for in $\Delta(j)$.

It can be concluded that for a system of two periodic elementary functions, "jumps" in terms of T_1 can occur for the feedback delay. Their occurrence depend on the characteristics of Δ, asynchronous vs. synchronous execution, and δ. A key characteristic is that the *age of data* is varying, i.e., delays associated with data-flow can occur. This is illustrated by the scenario where the time between successive sampling and actuation actions in the feedback loop is approximately constant, but where vacant sampling causes variations in the feedback delay. It should be noted that such data-flow oriented feedback delays are closely related to the loss of samples and can be interpreted as sensor noise, recall $v_\tau(k)$ in Figure 2.

4.3.4. Feedback Delay Variations from a Periodic to an Aperiodic Elementary Function

The case where EF_1 is periodic and EF_2 is aperiodic is considered. Since the functions are executed in sequence, vacant sampling is normally eliminated and the feedback delay can be expressed as

$$\tau_c(k) = \delta_1(k) - O_{prior,1}(k) + \delta_2(k) \qquad (16)$$

The delay, even though reduced on the average, is still varying according to Eq. (16). Whereas the sampling period of EF_1 will vary due to jitter, as $T \pm J_{max}$, the interval between consecutive starting triggers of EF_2 is in the following range: $[\delta_{min,1} + T_1 - \delta_{max,1}, \delta_{max,1} + T_1 - \delta_{min,1}]$. EF_2 thus represents a varying load. A sporadic "server" function is an alternative to a pure aperiodic function.

Sporadic "servers" handle aperiodic events while guaranteeing a bounded load. Let the period of the sporadic server be T_{ss}. If the interval between consecutive arrivals of samples is larger than or equal to T_{ss}, execution is triggered immediately as for the aperiodic function. If the interval is smaller than T_{ss} the enabling of the sporadic server should be delayed by $\{T_{ss} - [\delta_1(k + 1) + T_1 - \delta_1(k)]\}$ following sampling instant $(k + 1)$ of EF_1. Following instant k of EF_1, the instant of the subsequent enabling of the sporadic server, $t_{k,ss}$ can be written as follows:

$$Earliest\ t_{k,ss} = max\{t_k + \delta(k), t_{k-1,ss} + T_{ss}\} \qquad (17)$$

For the case where $T_{ss} = T_1$, and when the same global clock governs the execution of the functions, Eq. (17) reveals that once $\delta_1(k) = \delta_{max,1}$, the two functions will be synchronous with a phase, $\Delta = \delta_{max,1}$. I.e., $t_{k,ss} = t_k + \delta_{max,1} \rightarrow t_{k+1,ss} t_{k,ss} + T_1$. However, since the functions are triggered by the same clock it is much easier to use periodic functions in the first place.

For asynchronous execution, the variations should be reduced on average. However, once $\delta_{max,1}$ has occurred, an approximate phase can only be guaranteed a certain time due to the clock drift. Either the clock drift has the effect that T_{ss} is actually larger, or smaller than T_1. The former case means that more than one sample eventually will arrive during T_{ss} and before this happens the feedback delay will be slowly increasing. In the latter case vacant sampling may occur. This limits the applicability of the approach.

It can be concluded that the delay variations on the average can be reduced when the receiving function is aperiodic. A key characteristic is that the time between successive sampling and actuation actions in the feedback loop varies. This corresponds to a data- and control-flow delay in the feedback loop. The delay can be interpreted as a disturbance acting on the control signal, recall $w_\tau(k)$ in Figure 2.

4.4.   Time-Variations and Delays in Complex Systems

It is quite straightforward to generalize the presented equations for response and feedback delays to more complex systems. This basically entails summation over the delays intro- duced by a number of elementary functions. It is important that asynchronous periodic functions are modeled as separate functions in order to account for detection delays. This of course also relates to system functions such as communication, e.g., medium access protocols. One to many connections between elementary functions are common in control systems. This could for example be the transfer of reference values from a coordinating function to several servo functions. Delay differences are then of interest and can be derived based on the Equations presented in section 4.

As an example of a generalization, Eq. (18) gives the feedback delay for a feedback loop composed of n time-triggered periodic elementary functions where EF_1 performs sampling, EF_n performs actuation, and intermediate functions perform communication, computations, etc. The functions can be synchronous or asynchronous. The approximate relation follows due to the prior overheads included in $\tau_{sc,i}$.

It follows from Eq. (14) and (18) that the ratio of maximum to minimum delays can increase linearly with the number of functions in the system if suitable execution strategies are not employed. When the functions all have the same period T, the worst-case factor $(n-1)T$ appears in the feedback delay.

$$\tau_c(j) \sim \sum_{i=1}^{n-1}[\tau_{sc,i}(k(i))] + \delta_n(j), \text{ where } k(n-1) = j, k(n-2) = k''(k(n-1)), \text{ etc.}$$

(18)

The exact determination of timing characteristics requires detailed knowledge of the execution strategy and the predictability of the hardware and software components. In particular, elementary function characteristics in terms of durations of execution and communication must be well known.

The analysis of the effects of time-variations, however, becomes much more complicated for multi-rate multivariable control systems where elementary functions implement algorithms which have dynamics (i.e., use past states), and where non-trivial data-flow interactions take place within feedback loops as discussed in section 3. As a simple but relevant example, consider the internal feedback of the control signal from the control algorithm to the observer function in Figure 2. The observer algorithm is designed based on constant (or zero) delays of the output and control signals. Time-variations can occur for one or both of the signals, affecting the performance of the observer. The performance of the associated feedback loop(s) is in turn a function of the observer dynamics and all other functions in the loop, complicating the analysis.

5. Conclusions

The design and implementation of distributed real-time control systems poses a challenging multidisciplinary problem. Design parameters from involved fields provide a large solution space. In principle, timing problems can be handled either from the computer or control engineering side. Feedback delay and sampling period variations are however difficult to treat theoretically and the uncertainties adds to other problems in control design. As shown in section 4, time-variations can become very significant unless a suitable execution strategy is used.

The primary choice in design should thus be to eliminate relevant time-variations. There are then two rather obvious solutions at hand for the design of distributed real-time control applications. A "*dedicated solution*" is based on dedicated and predictable resources for every function and service, thus avoiding and limiting time-variations. The other solution is based on "*adequate multiplexing.*" It is fairly obvious that to arrive at constant feedback delays in this case, full control of triggering, scheduling and synchronization policies is necessary. An important part of the solution concept is the provision of a global time base that provides sufficient accuracy and precision for synchronization in the control application. A global clock drastically facilitates all aspects of distributed control system design, implementation and management.

For single rate systems, requirements on precisely time-triggered actions at I/O points and constant feedback delays can be translated into end-to-end deadlines for computations within the computer system. For multi-rate systems, however, timing requirements are also imposed on synchronization of threads of control and constant delays of data-flows within the computer system. This implies that application models must deal with entities/abstractions, such as feedback loops, that involve more than one period. The timing requirements on constant delays and synchronization have wider applicability than feedback control systems, for example, for handling mode changes in distributed systems. It can further be noted that the timing requirements of existing real-time process models constitute a subset of the proposed model and can be obtained by relaxing requirements, e.g., from constant delay to upper delay bound (reflected by deadlines and response time analysis).

It appears that there is no available scheduling approach for distributed systems implementation that is capable of dealing with non-trivial multi-rate control applications. Static scheduling is a natural solution in view of the presented timing requirements. A pure and system wide static scheduling solution may however not always be possible or preferable, e.g., due to a multi-vendor design case or because of a mixed application load. A mix of static and run-time scheduling appears natural.

The elimination of time-variations in the design of the computer system enables the use of standard control theory. To manage transient failures and unexpected timing failures it is still important that the sensitivity to data-loss and timing variations are investigated and that timing tolerances are derived. To compensate for data-loss it is natural to include prediction. This requires that the application is appropriately notified of communication failures.

Depending on particular design situations and economics, the system hardware architecture and resource management policies may be more or less fixed, thus constraining the allowable solution space. Alternative "non-optimal" solutions may then be required. Execution strategies can be modified to minimize the time-variations and control design can be attempted to provide robustness towards time-variations, as discussed in sections 2 and 4.

Further work is ongoing to bridge the gap from control design to real-time implementation. In a follow up work to this paper (Törngren et al., 1997) the derivation of timing requirements for different multi-rate system structures and transformations to computer science models are investigated. One important goal in further work is to provide a toolset that complements commercial computer aided engineering tools to support real-time implementation of control applications in distributed computer systems. Essential parts of such a toolset include additional modeling capabilities, and tools for control system structuring, allocation and scheduling.

Appendix: Notation

For discrete-time variables the following convention is used: $x_{type,i}(k)$, denotes x of a particular *type* for function/activity i at sampling instant k. If x is constant over time the time index is omitted.

$C_{min} \leq C(k) \leq C_{max}$ Uninterrupted execution time of an elementary function with bounds.

CLK   *Clock*

31

D	Deadline.
$\delta_{min} \leq \delta(k) \leq \delta_{max}$	Internal response time of an elementary function and its bounds (see Definition 8)
$\Delta(j), \Delta$	Time-varying (skew) and constant phase relations for periodic functions (see Definitions 5 and 6).
$0 \leq J(k) \leq J_{max}$	Sampling period jitter (see Definition 11).
$k^*(j), k''(j)$	Integers used to define skew and $\tau_{sc}(j)$, see Eq. (7) and Definition 12.
k, j	Integers.
LCM, GCD	Least common multiple and greatest common divisor of a set of periods.
$O_{prior}(k)$	Prior overhead, delaying the actual starting time of a function (see Definition 10).
$O_{during}(k)$	During overhead, overhead occuring during the execution of a function.
$p_{min}(j)$	Integer used in the definition of sensor to controlled delay, see Eq. (9).
$t_{actuate}(k), t_{sample}(k)$	Actual time when actuation and sampling occur respectively.
t_k	Nominal start of period time, giving by kT.
tol_T, tol_τ	Sampling period and feedback delay tolerances (see Eq. (1) and (2)).
T	Nominal (sampling) period for a periodic activity/function.
τ	Time delay in general.
$\tau_{detect}(j)$	Detection delay when an activity/function is not synchronized with an event.
$\tau_c(j), \tau_r(j)$	Feedback and response delay for a feedback-loop.
$\tau_{ca}(j), \tau_{sc}(j)$	Controller to actuator and Sensor to controller delay.

Acknowledgments

The author would like to thank Prof. Jan Wikander and Prof. Björn Wittenmark for providing inspiration and Jan Wikander and the anonymous referees for valuable comments on the structuring of the paper. Kristian Sandström and Christer Eriksson are acknowledged for valuable comments on section 3 in the paper. This work has been supported in part by the Swedish National Board for Industrial and Technical Development, project DICOSMOS, 93-3485.

References

Andreff, N. 1994. Robustness to jitter in real-time systems. Internal report, May 1994, Dept. of Automatic Control, Lund Inst. of Tech, Lund, Sweden. Doc No. ISRN LUTFD2/TFRT–5507-SE.
Åström, K., and Wittenmark, B. 1990. *Computer Controlled Systems, Theory and Design*. 2:nd edition, Prentice Hall.

Audsley, N., Burns, A., Davis, R., Tindell, K., and Wellings, A. 1995. Fixed priority pre-emptive scheduling. *J. Real-Time Systems* 8: 173–198.

Baker, T., and Shaw, A. 1989. The cyclic executive model and ADA, *J. Real-Time Systems* 1: 7–25.

Belle Isle, A. 1975. Stability of systems with nonlinear feedback through randomly time-varying delays. *IEEE Trans. Automatic Control* AC-20(1).

Berg, M., Amit, N., and Powell, J. 1988. Multirate digital control system design. *IEEE Trans. Automatic Control* 33(12).

Burns, A., and Wellings, A., 1994. HRT-HOOD: A structured design method for hard real-time systems. *J. of Real-Time Systems* 6(1).

Franklin, G., Powell, J., and Workman, M. 1990. *Digital Control of Dynamic Systems.* 2:nd edition. Addison-Wesley.

Godbout, L., Jordan, D., and Apostolakis, I. 1990. Closed-loop model for general multirate digital control systems. *IEE Proceedings* 137, Pt. D,(5): 329–336.

Halang, W. 1990. Simultaneous and predictable real-time control achieved by accurately timed computer peripherals. *Proc. 11th IFAC World Congress on Automatic Control* 7: 279–284.

Halang, W. 1992. Contemporary computers considered inappropriate for real-time control. *Proc. IFAC Algorithms and Architecturers for Real-Time Control* Pergamon Press.

Hatley, D., and Pirbhai, I. 1987. *Strategies for Real-Time System Specification.* New York: Dorset House Publ.

Hirai, K., and Satoh, Y. 1980. Stability of a system with a variable time delay. *IEEE Trans. on Automatic Control* AC-25(3): 552–554.

IEEE 1992. *The New IEEE Standard Dictionary of Electrical and Electronics Terms.* IEEE std. 100–1992. Fifth edition.

Jahanian, F., Lee, R., and Mok, A. 1988. Semantics of modechart in real time logic. *Proc. of 21sr Hawaii Int. Conf. on Systems Sciences.* pp. 479–489.

Klein, M., Lehoczky, J., and Rajkumar, R. 1994. Rate-monotonic analysis for real-time industrial computing. *IEEE Computer* January: 24–32.

Kopetz, H., Damm, A., Koza, C., Mulazzani, M., Schwabl, W., Senft, C., and Zainlinger, E. 1989. Distributed fault-tolerant real-time systems: The MARS approach, *IEEE Micro* 9(1): 25-40.

Kopetz, H., and Kim, K. 1990. Real-time temporal uncertainties in interactions among real-time objects. *Proc. 9th IEEE Symposium on Reliable and DIstributed Systems* Huntsville, AL.

Lauwereins, R., Engels, M., Ade, M., and Peperstraete, J. 1995. Grape-II: A system-level prototyping environment for DSP applications. *IEEE Computer* Feb.: 35–43.

Lawson, H. 1992. Engineering predictable real-time systems. *Real-Time Computing* NATO ASI series. Springer-Verlag.

Liu, C., and Layland, J. 1973. Scheduling Algorithms for multiprogramming in a hard-real-time environment. *Journal of the Association for Computing Machinery* 20: 46–61.

Luck, R. 1989. Observability and delay compensation of integrated communication and control systems. Ph.D. thesis, Dept. of Mechanical Engineering, Pennsylvania State University, U.S.A.

Mok, A. 1983. Fundamental Design Problems of Distributed Systems for the Hard Real-Time Environment. Ph.D. thesis, Massachusetts Inst. of Technology.

Motus, L., and Rodd, M. 1994. *Timing Analysis of Real-Time Software.* Pergamon.

Ray, A., and Halevi, Y. 1988. Integrated Communication and Control Systems: Part I—Analysis, and Part II—Design Considerations. *ASME Journal of Dynamic Systems, Measurements and Control* 110: 367–381.

Ray, A. 1994. Output feedback control under randomly varying distributed delays. *J of Guidance, Control and Dynamics* 17(4).

Shin, K., and Kim, H. 1992. Hard deadlines in real-time control systems. *Proc. IFAC Algorithms and Architectures for Real-Time Control* Seoul, Korea.

Shin, K., and Cui, X. 1996. Computing time delay and its effects on real-time control systems. *IEEE Trans. on Control Systems Technology* 3(2): 218–224.

Stankovic et al. 1995. Implications of classical scheduling results for real-time systems. *Computer* 28(6).

Stankovic, J., and Ramamritham, K. 1990. Editorial: What is predictability for real-time systems? *Real-Time Systems* 2(4): 247–254.

Tindell, K., and Clark, J. 1994. Holistic schedulability analysis for distributed hard real-time systems. *Microprocessing and Microprogramming* 40: 117–134.

Törngren, M. 1995. Modeling and Design of Distributed Real-time Control Applications. Ph.D. thesis, Dept. of Machine Design, The Royal Institute of Technology, Stockholm, Sweden.

Törngren, M., and Wikander, J. 1996. A decentralization methodology for real-time control applications. In the IFAC *J. Control Eng. Practice* 4(2), special section on the engineering of complex computer control systems. Pergamon Press.

Törngren, M., Eriksson, C., and Sandström, C. 1997. Deriving Timing Requirements and Constraints for Implementation of Multirate Control Applications. Internal report, 1997:1, Dept. of Machine Design, The Royal Institute of Technology, Stockholm, Sweden.

Uchida, K., and Shimemura, E. 1986. Closed-loop properties of the infinite-time linear-quadratic optimal regulator for systems with delays. *Int. J. Control* 43(3): 773–779.

Wang, Z., Lundström, P., and Skogestad, S. 1994. Representation of uncertain time delays in the H_∞ framework. *Int. J Control* 59(3): 627–638.

Voulgaris, P. 1994. Control of asynchronous sampled data systems, *IEEE Trans. on Automatic Control* 39(7)

Wittenmark, B., Nilsson, J., and Törngren, M. 1995. Timing problems in real-time control systems: Problem formulation. *Proc. of the American Control Conference* Seattle, Washington.

Xu, J., and Parnas, L. 1990. Scheduling processes with release times, deadlines, precedence and exclusion relations. *IEEE Trans. on Software Engineering* 16, 360–369.

Real-Time Systems, 14, 251–267 (1998)

Design and Programming Tools for Time Critical Applications

PAOLO ANCILOTTI
Scuola Superiore S. Anna Via Carducci, 40—Pisa, Italy   paolo@sssup.it

GIORGIO BUTTAZZO
Scuola Superiore S. Anna Via Carducci, 40—Pisa, Italy   giorgio@sssup.it

MARCO DI NATALE
Università di Pisa, Dip. Ingegneria dell'Informazione, via Diotisalvi, 3—Pisa, Italy

MARCO SPURI
Scuola Superiore S. Anna Via Carducci, 40—Pisa, Italy

Abstract. The development of time critical applications needs specific tools able to cope with both functional and non-functional requirements. In this paper we describe a design and programming environment to assist the development of hard real-time applications. An interactive graphic interface is provided to facilitate the design of the application according to three hierarchical levels. The development model we propose is based on an iterative process in which the real-time scheduling support is considered from the beginning of the design phases.

Our graphic environment integrates several tools to analyze, test, and simulate the real-time application under development. In particular, the tools we have implemented are: a Design Tool, to describe the structure of the application, a Schedulability Analyser Tool (SAT), to verify off-line the feasibility of the schedule of a critical task set, a Scheduling Simulator, to test the average behavior of the application, and a Maximum Execution Time (MET) estimator to bound the worst case duration of each task.

Keywords: development environment, real-time, scheduling, simulation, design tool

1. Introduction

In the development of Hard Real-Time (HRT) applications we have not only to verify the functional correctness of each task, but also to guarantee their temporal requirements in all workload conditions. If the results of a real-time computation are not produced in time, they are not just late, but wrong. Because of this, we usually experience an increase in the complexity of some of the last stages of the development, namely, verification and validation. However, this is also due to the weakness of the specification and design phases, which should explicitly support the analysis of the temporal characteristics as early as possible.

Some examples of development tools explicitly oriented to HRT applications are described in the literature. We have analyzed them by considering the abstractions provided to specify the temporal requirements of the applications and how these abstractions are integrated in the development method.

DARTS (Gomaa, 1984) is a development tool that allows to decompose the application in a set of concurrent processes. The requirements of the processes are specified through data-

35

flow diagrams. However, the specification of their temporal characteristics is not directly supported, and the goal is to achieve high performance through a good priority assignment.

An extension of the HOOD methodology for HRT applications is described in (Burns, 1991). Using this methodology, called HRT-HOOD, the designer can define a set of objects with attributes describing their temporal and dependability requirements. The development of the application is decomposed in two phases: the logical and the physical design. During the first phase, the requirements are transformed into objects, while in the physical phase the objects are allocated to the physical resources and the schedulability analysis is done. Owing to the adopted top-down approach, the development of HRT applications has the major drawback of reconsidering the whole design structure even if a temporal error is discovered in one of the later stages.

The methodology adopted in MARDS (Kopetz et al., 1991) is based on the MARS distributed system, in which the basic abstraction defined to decompose an application is the transaction, i.e., a computation that must be performed at specified intervals or as reaction to asynchronous events. Any transaction can be decomposed in subtransactions until the process level is reached. Schedulability and dependability analysis are done during any decomposition using estimated values. The aim is to get information on the feasibility of the application as soon as possible. However, the methodology heavily depends on the underlying architecture of MARS and has a top-down structure as the previous one.

The methodology proposed in PERTS (Liu et al., 1993) handles the real-time characteristics of an application at the design stage. The main abstractions used during the decomposition of an application are the task graph and the resource graph that contain the dependencies and the time attributes of the system objects. A fundamental tool of the system is the schedulability analyzer, that allows static verification of the time properties of the tasks and assists the user in choosing the right scheduling and resource allocation algorithms from an extensive internal library. Again, the real-time aspects of an application are considered during the design stages. Furthermore, in PERTS an iterative approach is proposed in order to allow the output of the schedulability analyzer to guide the allocation of tasks and resources and then to allow an early analysis of the time characteristics of the application. The specification and programming levels are essentially in cascade with the design stage. However, the measurement of the worst case computation times is not supported by a specific tool.

The main problem with the methodologies adopted in such environments (typical of cascaded approaches) is that a wrong choice in the first stages of the project may show its consequences only at later stages. In the worst case, an error encountered during the testing of the application can cause a backtrack of the development to the early design phases.

In IPTES (Pulli and Elmstrom, 1993 and Leon et al., 1993) the software production cycle is organized as an incremental goal. This permits to immediately verify the components of the application under development, leaving a coarser specification for the other components.

In the development model that we propose in this paper, we attempt to capture all the interesting aspects of these methodologies. Our model is thus based on an iterative development with rapid prototyping (Bohem and Papaccio, 1988 and Brooks, 1986), and the software production is done through a series of complete development cycles, from design to testing, where a cascaded approach is adopted in each cycle.

The timing requirements of the application and some characteristics of the computer architecture on which the application has to be implemented are considered from the early stages
of the production process. Moreover, the software development is explicitly supported
by dedicated tools which allow one to analyze, test, and simulate the process behavior in
different situations. A number of scheduling algorithms and protocols for accessing shared
resources have been developed for HRT systems. In our development tools we have considered several of these algorithms, which can be used to test the feasibility of the application
on different system architectures. Other non-functional requirements, such as dependability
and mode change, are not treated at the moment, even though the tools and the methodology
can be easily extended.

The proposed methodology for designing a real-time application consists of an iterative
execution of a conventional cascade model. Given three classes of activities with different
levels of criticalness, we believe that the most critical tasks should be separately developed and tested in the first iteration. Only when the core of the time critical application
has been consolidated, soft and non real-time tasks should be added. The advantage of
this incremental design is the early implementation of the critical modules that are more
influenced by the time characteristics of the architecture and by the choice of the system
algorithms.

The rest of the paper is organized as follows. Section 2 presents the structure of our
development environment, while the individual tools implemented are described in sections 3, 4, 5, and 6. Finally, conclusions and possible extensions to our work are discussed
in section 7.

2.   A Development Environment for Hard Real-Time Applications

The development environment proposed in this paper is composed of a set of tools arranged
in a cascade model to be cyclically utilized within the iterative approach described in the
previous section. A scheme of the proposed software development process is depicted
in Figure 1. The functionality of the design stages is implemented by three cooperating
tools: the DEsign Tool (DET), the Scheduling Analyzer Tool (SAT) and the scheduling
Simulator. The output of these tools goes to the programming tools: the definition of the
application tasks with their time attributes, and the description of the system resources go
into the Application Info file (containing the task names and their dependencies), necessary
to the MET tool for the estimation of the maximum execution times of the application
tasks. The dummy prototypes go to the compiler and from there to testing. The Class
templates are generated to give the programmers a coding framework consistent with the
design description.

The definition of the system design is supported by the Simulator and by the Scheduling Analyzer, that are designed to cooperate with the DET to give optimal control in the
definition of the system structure and algorithms. The MET tool works on the Application
Info File and the Application Code files and produces an accurate static estimate of the
maximum execution times, not only for the tasks, but also for the critical sections on shared
resources or, virtually, for any other portion of high-level code.

The result of the time estimates is fed back to the editors, for a possible refinement of

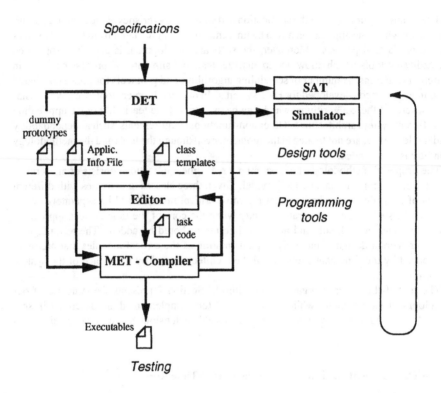

Figure 1. Software development process.

poorly efficient code, to the SAT, for a better testing of the worst case schedulability, and to the scheduling simulator, to improve the simulation of the average case behavior.

The executable code produced by the compilers should be extensively tested by the testing tools. These tools should allow an incremental testing for all the entities represented at the design level and later programmed, from tasks and resource instances to higher level components.

To timely trace the activity of the system during the testing of the dummy or actual modules, we have developed a monitoring tool, called TRACER, which is able to monitor the actual activation and execution times for all the tasks running in the system and give information about resource and channels usage. This tool has been implemented for the HARTIK system (Buttazzo and Di Natale, 1993 and Buttazzo, 1993), a hard real-time kernel for programming control applications with explicit time constraints and guaranteed execution, however it can easily be extended for other kernels.

A first version of the tools described in this paper has been used to design a real-time monitoring system dedicated to a vessel traffic control application, called TRACS, developed within the European Esprit Programme, project #6373 (Ancilotti et al., 1993). The aim of TRACS was to develop an advanced monitoring system able to integrate data from several radar stations and assist a human operator in controlling the navigation of vessels in a harbor environment.

In TRACS, periodic activities are devoted to the processing of radar images for extracting ship parameters like shape, area, position and speed. Other periodic tasks are dedicated to track the trajectories of the ships, compute their cross distance and detect dangerous situations for collision avoidance. Furthermore, the control system is designed to handle ship guidance, stranding avoidance, moored ships monitoring, and intrusion avoidance. Aperiodic tasks are associated to alarming events, which are generated when dangerous situations are detected. The nature of these situations and the main goal of avoiding accidents and especially loss of human lives, give rise to stringent deadlines for these tasks.

In the TRACS project, the design tool was intensively used to define the structure of the application and divide the activities into tasks. The MET estimator and the schedulability analyzer were very helpful during the implementation phase for producing a feasible schedule for the time critical tasks, thus reducing the cost of code refinement. The SAT has also been employed (using estimates) before the source code implementation for modifying some soft time constraints in order to balance the efficiency against the scheduling feasibility. Finally the scheduling simulator was used for testing the predictability of the system in the presence of high aperiodic load due to bursty arrivals of alarming events. The structure of the TRACS application is shown in Figures 3 and 4.

3.   The Design Tool

To facilitate the design of complex hard real-time applications, an interactive graphics tool allows the user to describe the application requirements according to three hierarchical levels: the node level, the component level, and the object level.

At the first level, the application is described as a number of nodes (which can be mapped to virtual or physical processors) that communicate through channels. Nodes and channels are graphically represented by icons linked with arrows. At the component level, the developer specifies the activities performed within a node, described as a set of concurrent tasks that communicate through shared critical sections or through channels. Parameters and attributes of tasks are defined at the object level.

Before the creation of any possible object (task, resource, channel, or message) a class for that type of object must be created. In the definition of a class all common characteristics that can be associated to a set of objects are specified. Figure 2 shows the window associated to the creation of a process class.

When defining a process class, the user has to specify the following fields:

- **Name:** represents the class identifier and can be any valid string of characters.

- **Periodicity:** describes the timing characteristic of the task class; possible choices are: *periodic, sporadic*, and *aperiodic*.

39

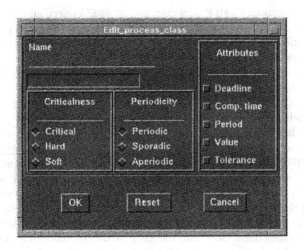

Figure 2. Example of process class creation.

- **Criticalness:** is related to the consequences that could occur if a task would miss its deadline; possible choices are: critical (a deadline miss could cause catastrophic consequences), hard (high damage), soft (low damage).

- **Attributes:** a list of attributes associated to each task belonging to this class; they are:

Deadline:	the latest finishing time of the task;
C_time:	the worst case computation time of the task;
Period:	the interval of time between two consecutive activations;
Value:	a positive integer which reflects the importance of the task;
Tolerance:	the maximum amount of time that a task is allowed to execute after its deadline.

Notice that the specific values of the attributes defined in a class are specified within each task instance in the object description level. The worst-case computation time for each task can be evaluated, once the source code is available, by a specific tool, described in section 6. At the beginning, only estimates can be used.

Like a task class, a resource class can be created from a proper menu by specifying the class name and a list of the operations that are allowed to manipulate a resource instance of this class. For each resource operation the user has to specify a name and a duration, i.e., the time necessary to execute that operation in the worst case.

A channel class defines all features that have to be associated with a set of channels. Examples of features that can be defined at this level include: synchronous channel, partially asynchronous channel with buffering capability, and totally asynchronous channel with non-consumable messages and overwrite semantics. Finally, a message class simply describes the type of data contained in a message.

40

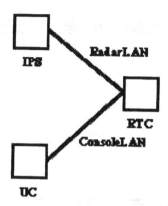

Figure 3. Design of the TRACS application at the node level.

Once an object class has been defined, an object instance can be created by opening a proper window. For example, to create a task, the user has to specify the following fields:

TaskName: a string of characters representing the task identifier;
TaskClass: the name of a task class (it can be selected from a class list);
Attributes: the values of the attributes specified in the task class.

Once all application components have been described, the user can easily connect nodes (and then tasks) through channels by a few mouse clicks. A shared resource can be created from a proper menu by specifying the resource name and the resource class. To associate a resource R with a process P, it is sufficient to click on the R icon, on the P icon, and to select the resource operations used by the process P from the operation list shown in the resource window.

Figure 3 shows an example of application design, at the node level, from the TRACS project (Ancilotti et al., 1993) described in the previous section. In TRACS the sensory data coming from multiple radars are merged and sent through a local area network to a real-time controller where dangerous or forbidden conditions are detected. Positions of the ships, together with the alarms and warnings, are sent through a local area network to another station containing the user console.

The application consists of three nodes connected by logical channels: the Image Processing System (IPS), the Real-Time Console (RTC) and the User Console (UC). IPS is the node containing the radar data acquisition processes, RTC is the node containing the real-time controller and the constraint checkers, and the UC node represents the user console.

Messages associated with the channels, together with their timing properties and constraints (if any), are described at this level. Figure 4 represents the internal structure of the RTC node at the component level, showing tasks (circles), resources (gray squares) and channel ports (black squares). The channel from the IPS (RadarLAN) is accessed through the corresponding port, and *mixInput* is the process that manages the connection. The

41

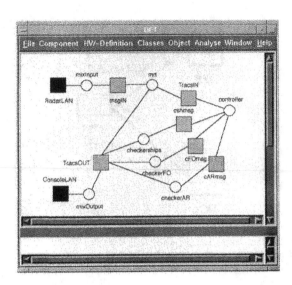

Figure 4. Design of an application at the component level.

periodic process *mrt* receives the messages containing the positions of the ships from the mailbox *msgIN* and sends one message to the *controller* process and another to the output mailbox. The controller dispatches data to the actual controllers, i.e., the processes *checkerships*, *checkerFO* and *checkerAR*, which can send alarms and warnings to *mixOutput*. This last process simply passes the incoming messages to the *ConsoleLAN* port.

The output of the design tool goes to the programming level and to the scheduling tools. At the programming level, the developers can make use of the class templates containing the declarations and the skeleton of the program tasks together with the corresponding communication declarations and primitives, as specified at the design level.

Once the development of the code is completed, the application files are examined by another tool: the Maximum Execution Time (MET) estimator that analyzes the code of each task and returns the estimates of the worst case computation times.

4. Schedulability Analyzer

The Schedulability Analyzer Tool (SAT) is very useful for designing predictable real-time applications, because it enables the developer to analyze a set of tasks and statically verify their schedulability. In a hard real-time system a schedule is said to be feasible if all critical tasks can complete within their deadlines.

If the schedulability analysis gives a negative result, the user can change task parameters and rerun the guarantee test. For instance, some adjustments are possible by modifying the task deadlines, or by producing a more compact and efficient code for some critical tasks. Depending on the application needs, when the complete set of tasks is found unschedulable, the user could also accept the task set anyhow, since the schedulability analysis is done assuming worst case computation times. In this case, the user can indicate the importance level of each task, so that if a transient overload occurs during execution, only the least value tasks will be rejected.

When selecting the schedulability analysis from the main menu, the window shown in Figure 5 appears on the screen. It contains the following fields:

- **Selection:** this field allows the user to select the tasks to be analysed by the guarantee algorithm. Two options are available: by name and by criticalness.

- **Processes:** this field becomes active when choosing the selection by name, and it displays the list of all the processes defined in the node under consideration.

- **Criticalness:** this field becomes active when choosing the selection by criticalness, and it displays the list of the critical attributes for the processes defined in the current node.

- **Scheduling Info:** this field displays the result of the scheduling analysis as a *Yes* or *No* answer. When an only sufficient condition is not satisfied, the string '*??*' is displayed. The total utilization factor is also shown in this window, and it takes into account the periodic load, the sporadic tasks with minimum interarrival time, and aperiodic servers.

- **Scheduling algorithm:** this field allows the user to select the scheduling algorithm to be used in the current node for periodic processes. The guarantee tests we have implemented are for the following scheduling algorithms: Rate Monotonic (Liu and Layland, 1973), Deadline Monotonic (Audsley et al., 1992), Earliest Deadline First (Liu and Layland, 1973), and Least Laxity First (Mok, 1983). Depending on the selected algorithm, the tool also allows the user to choose a server mechanism for soft aperiodic service (Sprunt, Sha and Lehoczky, 1989 and Spuri and Buttazzo, 1996).

- **Resource protocol:** this field allows the user to select the access protocol to be used among processes that communicate through shared resources. The possible protocols depend on the selected scheduling algorithm. For example, with the Rate Monotonic algorithm we can choose the Priority Inheritance (Sha, Rajkumar and Lehoczky, 1990) or the Priority Ceiling protocol (Sha, Rajkumar and Lehoczky, 1990), whereas with the Earliest Deadline First algorithm we can choose the Dynamic Priority Ceiling (Chen and Lin, 1990) and the Stack Resource Policy (Baker, 1991).

The overhead due to the execution of the timer interrupt routine can also be taken into account by specifying the worst case execution time of the timer routine and the system tick, i.e., the time quantum which defines the system clock resolution.

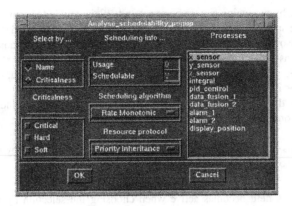

Figure 5. Schedulability analysis window of a node.

5. Scheduling Simulator

Many practical real-time applications do not contain critical activities, but only tasks with soft time constraints, where a deadline miss does not cause any serious damage. In these applications the user may be interested in evaluating the performance of the system in the average case behavior, rather than in the worst case behavior.

In order to do that, a statistical analysis through graphic simulation is required. For this reason, our development environment includes a scheduling simulator and a load generator tool for creating random aperiodic activities. Actual computation times, arrival times, duration and position of critical sections in the tasks are computed by the load generator as random variables, whose distribution is provided by the user in a separate window.

Once all simulation parameters have been provided, the scheduling simulation can be started by clicking on the corresponding button from the main menu. The simulation can be observed in a time interval, called *observation window*, equal to the least common multiple of the periods of all periodic tasks. If the observation period is less than 400 μsec, the time scale is automatically chosen so that the scheduling is plotted in the full window. If the observation period is greater than 400 μsec, then the simulator chooses a time scale such that one pixel corresponds to 1 μsec. If the observation period is greater than 800 μsec, the whole scheduling does not fit in the window. In this case, the buttons ">>>", "<<<", or the button "**Unzoom** ...", can be used to change the observation interval and show the rest of the scheduling. Figure 6 shows an example of a scheduling simulation under Rate Monotonic.

Task executions are represented by rectangles, task arrivals by up arrows, and task deadlines by down arrows. Periodic deadlines coincident to the end of the period are indicated with a vertical segment with no arrow. A time overflow is marked by a circle on the missed deadline. Dark areas within task execution represent access to critical sections or monitor procedures.

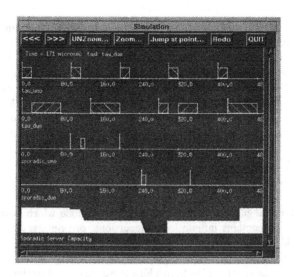

Figure 6. Example of a simulated schedule under Rate Monotonic.

By one click of the mouse pointer on the task execution, the simulator displays the time instant in that point, the name of the selected procedure (*main* means non critical code) and, in the case of nested procedures, the names of the calling procedures. Finally, the simulator detects a global deadlock and displays the time instant at which it occurs.

Our simulator has been designed to simulate different types of server mechanisms for soft aperiodic scheduling. Server parameters must be specified by the user within a proper subwindow, opened from the *Options* menu. These parameters include the server period T_s, the server capacity C_s, and a list of aperiodic tasks associated to the server. When a Sporadic Server is defined, it is convenient to start the schedulability test to verify if the task set is feasibly schedulable. If the set is not schedulable, it is necessary to reduce C_s or increase T_s, to make it feasible. The server capacity graph is also represented in the simulation and, using the mouse pointer, it is possible to know its value in each point.

6. MET Estimator

The construction and verification of predictable hard real-time systems need an accurate study of the time characteristics of the application tasks. To determine whether a critical task will complete before its deadline, its computation time must be bounded and its worst case execution time should be evaluated. Given the maximum execution time for each task, the scheduling algorithm is able to check the schedulability of the whole set of tasks.

At present, real-time application designers run programs with test data, in order to have an estimate of the execution time of the application tasks. This process is fault prone and

45

requires a lot of additional work for testing and debugging. Note that run time determination of maximum execution time (MET) is very difficult to pursue, since the control flow of a program usually depends on input data and current variable settings. In practice, it is impossible to simulate program executions for all possible variable settings. On the other hand, a static analysis must be supported by a programming style and specific language constructs to get analysable programs. The problem of estimating the MET of a task is indeed quite complex, since it depends on several factors, such as processor type and speed, hardware architecture, operating system, compiler and language constructs.

6.1. Language Extensions

Most programming languages allow the development of code which is intrinsically un-bounded, because it contains infinite loops, recursions, or primitives which may cause unbounded blocking.

To write time-bounded code, we have extended the C language with a small set of con-structs. A monitor-like construct has been added to isolate critical sections and to evaluate task blockings while accessing shared resources. It is possible to program optional bounds in order to limit the number of iterations in loop statements or to limit the number of pro-cessed conditional branches inside loops. We have modified the syntax of some C constructs as in the example shown in Figure 7.

The meaning of max(100) in the while construct is that the loop can be executed at most 100 times. In the if construct, max(20;90) means that (a[i].nice) is true at most 20 times and is false at most 90 times within the scope of the closest nested loop. When the programmer does not put any bound on the loop constructs in the coding phase, the tool asks for them at evaluation time. The critical sections can be easily expressed in terms of a monitor-like construct, as outlined in Figure 8.

Note that the construct is similar to the monitor introduced in (Hansen, 1975), and used to share data structures among several tasks. The extended constructs are recognized by a precompiler that we have developed. The source code written in our extended language is transformed into standard C code. In this phase, detailed information about tasks and critical sections are collected in order to develop a complete description of the time behavior of the application.

6.2. MET Evaluation

The hardware factors that influence the MET are the processor clock, the number of CPU (or coprocessor) cycles taken for executing every instruction, the efficiency of the processor pipeline and of the cache, and the number of wait states when accessing the bus for memory or I/O operations. All these factors should be considered in the software model of the architecture (Niehaus, Nahum and Stankovic, 1991 and Harmon, 1990).

The model we use includes the simulation of the processor in a table-driven fashion, where assembly instructions are translated into execution times depending on their operating code,

```
while (i<=n) max(100) {
     if (a[i].nice) max(20;90)
          sum += a[i].weight;
     i++;
}
```

Figure 7. Example of extended while and if constructs.

```
monitor <monitor_name> {
     <private data and procedures>
public:
     <access procedures>
}
```

Figure 8. Example of shared data structure management.

operands and addressing mode. At the moment, we assume that all other components of the system (e.g., memory, bus, cache, processor pipeline, i/o) have an ideal timing behavior.

The MET evaluation tool we have developed is conceived to work as illustrated in Figure 9. The Application Info File containing information about the application tasks (names of the source files, dependencies, resource requirements) is one of the inputs. The other input is the set of the Application code files containing the actual program. The program files are opened and analyzed by a precompiler, all non-standard constructs are removed, and the resulting code (now standard C code) is compiled using a commercial C compiler. The user is prompted for input any time the information on the maximum number of iterations in the loops is missing. The information relative to loops and conditional executions is stored in a different file containing the semantic information and the description of the dependencies among tasks and resources. The compiler produces an assembly listing of the code together with the object files of the application.

The assembly code is then interpreted by the tool, which builds an intermediate graph representation of the program's control structure in terms of temporal behavior. A weight

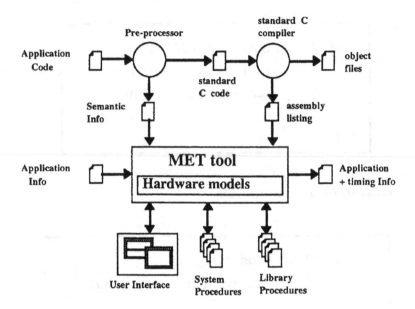

Figure 9. General scheme of the MET evaluation tool.

is assigned to every branch of the graph, corresponding to the number of CPU cycles that are needed for its execution. The graph is visited to evaluate the maximum cost path in the graph. The tool recognizes loops, starting from the deepest nested level, and reduces them to single nodes using the information about the maximum number of iterations.

After this stage, automatically executed for every procedure (starting from those with no dependencies or dependent only on library or system code), the tool produces an estimate of the maximum execution time for each task, expressed in number of CPU cycles. Once the system processor and the clock speed are known, the CPU cycles spent in the execution of the code can be translated into actual execution time (given our approximations on caches and pipelines).

The final estimated values are used to update the Application Info File, used as a data exchange structure with the Design Tool. The execution time relative to system calls and standard library procedures are evaluated once for all and stored in a table. The MET tool has been implemented on X Windows and produces a display as shown in Figure 10.

The lower section of the window shows the project the tool is working on, and the application files that are currently being evaluated. The upper section of the window shows the C file containing the application procedures while the middle section contains the corresponding assembly code generated by the compiler. The user loads the Application Info file into the tool, then, automatically, the tool loads the code files and analyzes the procedures it finds in the C-code, asking for the maximum number of iterations in loops (showing the first line of the loop in the window).

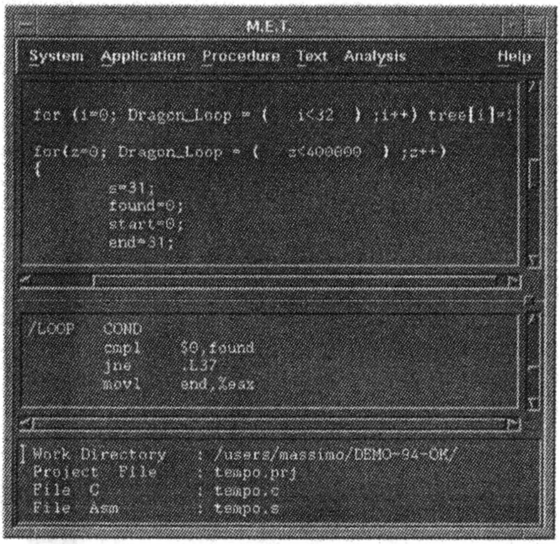

Figure 10. The MET estimator tool interface.

The menu commands allow the user to change the processor tables or to examine specific task procedures or monitor resources. The user can, for example, examine the control graph of any application procedure with the corresponding execution cycle for each segment, or evaluate the time spent in executing the monitor procedures. Figure 11 shows an example of the representation of an execution graph.

7. Conclusions

In this paper we have described a graphic environment for assisting the design and the programming of hard real-time applications. Our environment facilitates the development of complex hard real-time applications and allows the user to describe the application according to three hierarchical levels: the node level, the component level, and the object level.

With respect to other approaches proposed in the literature, our design cycle is based on an extended spiral meta-model in which the real-time scheduling support is considered from the beginning of the design process. One advantage of this approach is to drastically reduce the number of trial-and-error iterations, by examining the non-functional requirements of the applications in the early design phases.

A first version of the development environment has been used to design a real-time application dedicated to a vessel traffic control system developed within the ESPRIT project TRACS (Ancilotti et al., 1993). Currently, we are considering several improvements of the tools available in our integrated environment, such as a more precise evaluation of the worst

49

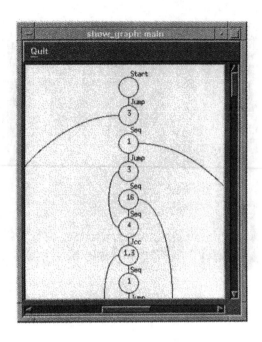

Figure 11. Example of procedure graph.

case computation time, a larger number of algorithms and resource policies supported by the Schedulability Analyzer and by the Simulator and, finally, a more flexible user interface.

Acknowledgments

This work has been supported in parts by MURST 40% and by the CNR of Italy.

References

Ancilotti, P., Buttazzo, G. C., Di Natale, M., and Spuri, M. 1993. TRACS: A flexible real-time environment for traffic control systems. *Proceedings of IEEE Workshop on Real-Time Applications* New York, NY, pp. 50-53.
Audsley, N., Burns, A., Richardson, M., and Wellings, A. 1992. Hard real-time scheduling: The deadline monotonic approach. *IEEE Workshop on Real-Time Operating Systems.*
Baker, T. P. 1991. Stack-based scheduling of real-time processes. *The Journal of Real-Time Systems* 3(1): 67–100.
Bohem, B. W., and Papaccio, P. N. 1988. Understanding and controlling software costs. *IEEE Transactions on Software Engineering* 15(7): 902–916.
Brooks, R. A. 1986. A robust layered control system for a mobile robot. *IEEE Journal of Robotics and Automation* 2(1): 14–23.
Burns, A. 1991. Scheduling hard real-time systems: a review. *Software Engineering Journal* May: 116–128.

Buttazzo, G. C., Di Natale, M. 1993. HARTIK: a hard real-time kernel for programming robot tasks with explicit time constraints and guaranteed execution. *Proceedings of IEEE International Conference on Robotics and Automation* Atlanta.

Buttazzo, G. C. 1993. HARTIK: a hard real-time kernel for robotic applications. *Proceedings of IEEE Real-Time Systems Symposium*, Raleigh-Durham.

Chen, M., and Lin, K. 1990. Dynamic priority ceilings: A concurrency control protocol for real-time systems. *Real-Time Systems* 2.

Gomaa, H. 1984. A software design method for real-time systems. *Communications of then ACM* 9(27): 938–949.

Hansen, P. B. 1975. The programming language concurrent pascal. *IEEE Transactions on Software Engineering* SE-1(2): 199–207.

Harmon, M. 1990. Predicting execution time of real-time programs on contemporary machines. *Proceedings of the 28th Annual Southeast Regional Conference* ACM, Greenville S. Carolina.

Kopetz, H., Zainlinger, R., Fohler, G., Kantz, H., Puschenr, P., Scutz, W. 1991. The design of real-time systems: from specification to implementation and verification. *Software Engineering Journal* May: 72–82.

Leon, G., Due Nas, J. VC., De La Puente, J. A., Alonso, A., Zakhama, N. 1993. The IPTRES environment: support for incremental etherogeneous and distributeds prototipyng. *The Journal of Real-Time Systems* 5: 153–171.

Liu, C. L., and Layland, J. W. 1973. Scheduling algorithms for multiprogramming in a hard real-time environment. *Journal of ACM* 20(1): 46–61.

Liu, J. W. S., et al. 1993. PERTS: A prototyping environment for real-time systems. *Proceedings of IEEE Real-Time Systems Symposium* Raleigh-Durham.

Mok, A. K. 1983. Fundamental design problems of distributed systems for hard real-time environments. Ph.D. Dissertation, MIT.

Niehaus, D., Nahum, E., and Stankovic, J. A. 1991. Predictable real-time caching in the spring system. *IFAC Real-Time Programming*: 79–83.

Pulli, P., Elmstrom, R. 1993. IPTES: A concurrent engineering approach for real-time software development. *Real-Time Systems* 5:139–152.

Sha, L., Rajkumar, R., and Lehoczky, J. P. 1990. Priority inheritance protocol: An approach to real-time synchronization. *IEEE Transactions on Computer* 39(9): 1175–1185.

Sprunt, B., Sha, L., and Lehoczky, J. P. 1989. Aperiodic task scheduling for hard real-time systems. *Real-Time Systems* 1: 27–60.

Spuri, M., and Buttazzo, G. C. 1996. Scheduling aperiodic tasks in dynamic priority systems. *Real-Time Systems* 10(2): 179–210.

Real-Time Systems, 14, 269–291 (1998)
© 1998 Kluwer Academic Publishers, Boston.

Rapid Prototyping of Real-Time Information Processing Units for Mechatronic Systems

H.-J. HERPEL   hans.juergen.herpel@dss.dornier.dasa.de
Dornier Satellitensysteme GmbH, D-88039 Friedrichshafen,

M. GLESNER
Darmstadt University of Technology, Institute for Microelectronic Systems, Karlstr. 15, D-64283 Darmstadt

Abstract. Mechatronics is a rapidly growing field that requires application specific hardware/software solutions for complex information processing at very low power and area consumption, and cost. Rapid prototyping is a proven method to check a design against its requirements during early design phases and thus shorten the overall design cycle. Rapid Prototyping of real-time information processing units in mechatronics applications requires code generation and hardware synthesis tools for a fast and efficient search in the design space. In this paper we present a rapid prototyping environment that supports the designer of application specific embedded controllers during the requirement's validation phase.

Keywords: rapid prototyping, real-time systems, mechatronic, hardware-software co-design, ASIC synthesis, ASIC emulation

1. Introduction

The quality of a product heavily depends on the quality of the process that creates it. In mechatronics system design many people from many different engineering disciplines are involved in this creation process. Every engineering discipline has its own language, models and culture. This makes communication between customers and developers and also among developers from different disciplines difficult and inefficient.

Pei Hsia et al. (1993) pointed out, that without a well-written requirement's specification, developers do not know what to build, customers do not know what to expect, and there is no way to validate that the system as built satisfies the requirements.

Requirements engineering is the disciplined application of proven principles, methods, tools and notations to describe a proposed system's intended behavior and its associated constraints. The primary output of requirements engineering is a requirement's specification. It must treat the system as a black box. It must delineate inputs, outputs, the functional requirements that show external behavior in terms of input, output, and their relationships, and nonfunctional requirements and their constraints, including performance, reliability and safety.

As proven by many authors (e.g. Boehm, Gray and Seewaldt, 1984; Ketabchi, 1988; Gomaa, 1990), *rapid prototyping* is a promising approach in requirements engineering. It means the construction of an executable system model to enhance understanding of the problem and identify appropriate and feasible external behaviors for possible solutions. The model is executed in real-time, thus, tuning parameters of an algorithm is much faster

53

compared to conventional simulation. Prototyping is an effective method to reduce risk on mechatronics projects by early focus on feasibility analysis, identification of real requirements, and elimination of unnecessary requirements. Rapid prototyping in mechatronics system design includes functional, performance, and user interface aspects (Herpel, Wehn and Glesner, 1995). The customer gets a system to play around with very early in the design cycle. This helps to bridge the previously mentioned communication gap and to create a common basis of understanding.

The following sections describe our approach to Rapid Prototyping in the field of mechatronics. Based on the characteristics of our application's domain (section 2) a design methodology is derived (section 3). The methodology comprises a system model to describe the application, a life cycle model that defines the development steps, and a set of tools to support the system designer. Section 4 outlines the basic design steps from system description down to prototype synthesis. Main focus here is on hardware generation. In section 5 two mechatronics projects are presented.

2. System Characteristics

The domain dependency of computer-aided design systems grows with the level of abstraction. Therefore, many mechatronic projects were analyzed carefully before a design methodology was defined. The system characteristics as listed below are based on this analysis. The system characteristics of the information processing units (IPU) in mechatronics applications can be summarized as follows:

- The IPU of mechatronic systems is exogen driven. It interacts with the mechanical components but in most cases not directly with the user of the system.

- The IPU is triggered externally by periodic and non-periodic events. Mechanical processes are continuous in time and value.

- The IPU itself is discrete in time and value.

- The correctness of the system response depends not only on the correctness of the transformation but also on the time when the result is presented to the environment (real-time system).

- The requirements on data throughput and response times are very high (typically less than 100 μs response time).

- The IPUs have only a few number of subfunctions (less than 10 in the investigated systems).

- The IPUs of the investigated systems need to be installed very close to the mechanical components (no distributed systems).

- The electronic component of a mechatronic system must be controllable, time invariant and deterministic.

- The requirements on reliability and safety are very high, since the system does not operate under human supervision (embedded system).

- The IPU is heterogeneous and application specific in various aspects. It contains dataflow and controlflow oriented functions, and analog and digital circuits. The demand for customer specific adaptation requires software functions. Other subfunctions have to be realized in hardware because of performance and speed requirements or requirements on low power consumption.

3. Design Methodology

A methodology based on rapid prototyping was developed within the framework of the basic research program "Integrated Mechano-Electronic Systems (IMES)" at Darmstadt University of Technology (Herpel, Held and Glesner, 1994). The main goal was to bridge the previously mentioned communication gap between customers (engineers from the mechanical engineering department) and developers (electronic engineers and computer scientists) through requirement's animation and thus shorten the overall development cycle. The methodology called MCEMS[1] consists of a life cycle model, a system description model and a set of tools for prototype generation.

3.1. System Model

Any design problem begins with the characterization of the environment in which the system is expected to operate and a goal. The developer's job is to produce a design for a system that, when inserted in the specified environment, achieves the specified goal.

An abstract model is required to describe both the intended system and the (existing) environment. Systems theory offers some basic principles at a high level of abstraction: *features, active elements* and *networks*.

Features are used to represent time-varying properties of the environment. It takes on a single value at each point in time. The set of all possible values for a feature is called its *domain*. Active elements are used to represent active entities that sense and control features. A network is a composite object consisting of interconnected features and active elements. Whenever there is a connection between a feature and an active element, the feature is called *input feature* of the active element, and the active element is called *consumer* of the feature. Whenever there is a connection between an active element and a feature, the active element is called *supplier* of the feature and the feature is called an *output feature* of the active element. A network is *closed*, if and only if every feature is the output of at least one constituent active element; otherwise, we say the network is *open*. A network is *coherent* if and only if every feature is the output of at most one constituent active element; otherwise, we say the network is *incoherent*. A network is *self-loopless* if and only if a feature is not input and output feature of the same active element; otherwise, we say the network has *self-loops*.

In mechatronics typical domains are material, energy, or information. Accordingly, active elements can be divided into three classes (Färber, 1992):

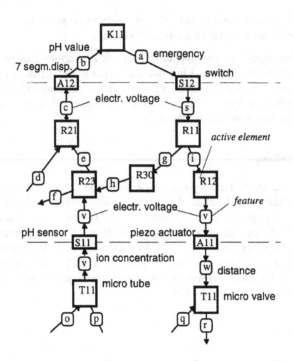

Figure 1. System model.

- **Technical active elements:** A technical active element is an element that transforms, processes, or transports material or energy. In accordance with DIN[2] 66201 we will use the expression *technical process* or simply *process*.

- **Computational active elements:** An active element is called computational active element, if information is processed. This includes the execution of a program on a computer (*task*) and also hardware components like a processor, for example.

- **Cognitive active elements:** Many technical and computational active elements do not run without human interaction. The user of technical/computational active elements picks up the information, processes it and transforms it into actions. This kind of active element is called cognitive process.

An active element performs its transformation either cyclically or when triggered by one of its inputs.

Now, a mechatronic system can be modeled as a coherent and directed network together with a compatible assignment for the domains of all features (Fig. 1). This compatibility restriction here is necessary to avoid nonsensical systems of various sorts. The network

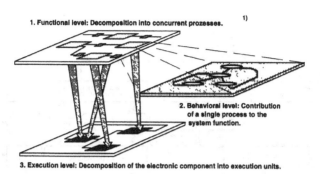

1. Functional level: Decomposition into concurrent prozesses.   1)

2. Behavioral level: Contribution of a single process to the system function.

3. Execution level: Decomposition of the electronic component into execution units.

1) Process = Transformation and/or transport of material, energy and/or information (DIN 66201)

Figure 2. Three views of the system.

may contain loops if there is at least one triggered active element in the loop. If a self-loop is closed over a triggered active element, the network may also contain self-loops.

All tasks together represent the information processing unit of a mechatronic system. The technical and cognitive processes build the environment in which the information processing unit is supposed to operate.

The conceptual model we developed for the information processing unit and its environment is based on the previously described system model. It includes three views of the system: *functional* view, *behavioral* view, and *execution* view (Fig. 2).

The *functional view* describes the proposed system as a collection of interconnected concurrent active elements (*process structure diagram*). This view is completely technology independent.

The *behavioral view* describes the contribution of a single process to the overall system behavior. A single model is not sufficient to describe technical and cognitive processes, and tasks. The models we use to describe the behavior of active elements include simple functional relations between input and output features, look-up tables, finite state machines, fuzzy rule base and neural nets. Different models can be combined within the behavioral description of a single active element or at the network level.

The *execution view* describes the information processing part of the mechatronic system as collection of hardware objects (*execution structure diagram*). This includes signal conditioning units, data conversion units, processor units, etc.

3.2. Life Cycle Model

Several steps are necessary to validate the user's requirements at each level of description before the actual system is being produced. The basic design steps in MCEMS (Fig. 3) are a combination of the rapid prototyping approaches used in software engineering (Lugi, 1988) and ASIC emulation (Walters, 1991).

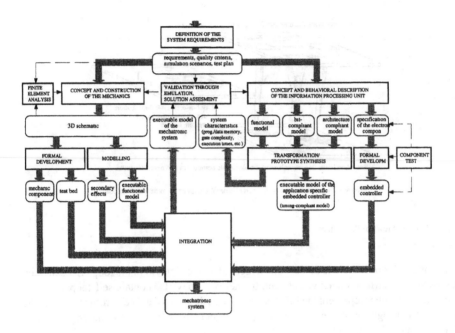

Figure 3. Life cycle model.

The rectangles in Fig. 3 show the activities and the rectangles with rounded corners represent the results of an activity. In the first step the mechatronic system is split into the embedded controller and the environment in which the controller is supposed to operate. The left path in Fig. 3 describes the (simplified) process to create a mechanical component. The right path shows the steps from formal specification to an executable model of the embedded controller.

Several levels of detail are necessary to completely model the application specific embedded controller and check it against its requirements (Adé et al., 1993):

At the **functional-compliant** level the correctness of the chosen algorithm is tested, and algorithmic parameters are tuned.

The **bit-compliant** model performs all computing with the same word-length as the final implementation. At this level all finite word-length effects (rounding, truncation, overflow/underflow) are analyzed. Its main goal is to reduce the word-length as much as possible and thus, save silicon area, and cost.

The **architecture-compliant** model uses the same architectural entities as the final implementation. At this level a trade-off between different architectures can be performed and communication mechanisms can be tested.

The central path in Fig. 4 shows the validation step. The model of the environment and the model of the embedded controller are integrated and checked against the user's

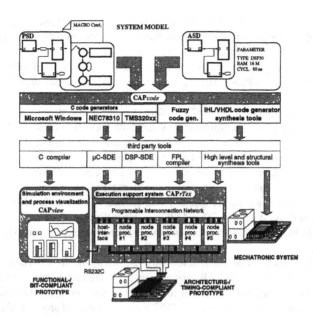

Figure 4. System configuration.

requirements. The model of the environment is treated as a black box. Since the right and left side of the life cycle model are decoupled, the model of the environment can be either a simple mathematical model, a test bed or the real hardware. However, it determines the timing. Therefore, the model of the embedded controller must always be **timing-compliant** with the actual implementation of the environment.

3.3. Supporting Tools

A set of software tools (**CAP***tools*) supports MCEMS. This covers formal entry of require-ments, mapping of the functional structure and behavioral description onto the execution structure, and interactive simulation and graphical animation.

The system model can be executed either on the development computer itself or in real-time with real components connected to the system model (hardware-in-the-loop simula-tion).

Fig. 4 gives an overview on all elements of **CAP***tools*. A graphical editor (**CAP***edit*[3]) is used to enter the functional and non-functional requirements as process structure diagram and to describe the execution system as execution structure diagram.

The behavior of the individual processes is specified either directly through templates or through import of C functions generated by other tools (e.g. Statemate, MATRIX$_x$). Application specific elements can be combined with elements for interactive input and output to build complete simulation scenarios. Actually, 22 library elements are available to create soft-panels, to read/store data from/into files, to generate animated process diagrams, and to communicate with external devices (e.g. **CAP***rTex*).

After compilation the process structure graph can be executed on the development computer under **CAP***view*[4]. **CAP***view* is a program for interactive simulation of process structure diagrams.

A real-time execution system (**CAP***rTex*[5]) allows to execute a system model in real-time. **CAP***rTex* is a heterogeneous multiprocessor system with six node processors (Herpel, Held and Glesner, 1993). The system topology (linear, ring, tree, . . .) is software programmable.

The heart of the programmable backplane are special interconnection chips (**IMES-***CP*516) with five bi-directional 16 bit ports. Each port can be connected to any other port except itself. The internal logic allows one-to-one and one-to-many connections as well as multiple paths at a time. A second ASIC (**IMES-***ICM*) was developed to support the interprocessor communication. This ASIC combines different data buffering (dual-port RAM, FIFO, ring buffer) and synchronization mechanisms (semaphore, interrupt, mailbox) on a single chip.

Several standard processor boards (microcontroller, DSPs) and an ASIC emulation board are available as node processors. All boards have the same architecture: a motherboard provides program memory and all services for interprocessor communication; the plugged-in data processing unit (DPU) consists of a CPU, local memory, and analog and digital I/Os. The DPU of the ASIC emulation node includes five SRAM based FPGAs (Xilinx XC4005) which provide a maximum capacity of 25.000 gate equivalents. These FPGAs are grouped around programmable interconnection chips (**IMES-***CP*516) to have more flexibility for interconnecting the FPGAs. This leads to a better logic utilization factor than fixed numbers of interconnection lines.

An *execution structure diagram* defines the configuration of the execution support system (number, type, and location of the node processors, and its interconnection structure). **CAP***edit* provides library elements for the actual implementation of the execution system **CAP***rTex*. The implementation details of a node processor are described by a set of attributes. This includes type of processor, cycle time, size of program, data, and communication memory, characteristics of AD and DA converters, for example. A set of synthesis tools supports the mapping of the *process structure diagram* onto the execution support system. The actual implementation of the mapping and code generation tool (**CAP***code*[6]) supports only manual assignment of processes to executable elements. The code for the different standard and application specific processors as well as the configuration bitstream for the execution system is generated by the tools.

4. Design Process

4.1. System Concept

The first step in the design process is to decompose the system into the embedded controller and its environment, and to define the interfaces between the two blocks. Also, the simulation scenario must be defined to test the system under development against the user's requirements. Next, the embedded controller must be further refined at the level of concurrent tasks. Since we have to deal with several periodic and non-periodic events, we defined the following steps for the decomposition:

1. Indentify all events in the system.

2. Define an object (a task) per event that handles the event.

3. Determine the import features of all tasks.

4. Determine the export features of all tasks.

5. Determine the object attributes.

6. Connect all objects according to their data dependencies.

The next step in the design process is to describe the behavior of the individual tasks. FSMs, high level programming languages, tables, and fuzzy rule bases can be used to describe this behavior (either functional or bit-compliant). Next, the model is converted into an executable model and integrated with the model of the environment where the embedded controller is supposed to operate. The simulation environment (**CAP***view*) we developed for that purpose runs under Microsoft® Windows™ and provides a comfortable way to interact with the simulation models.

During algorithm development it is common practice to use floating-point arithmetic. This helps to concentrate on the algorithm itself rather than to take care on rounding and truncation effects. But even with 0.8 μm CMOS technology it is costly to implement floating-point arithmetic on a chip. Thus, the conversion into a solution that uses fixed-point arithmetic can help to save Si-area and cost. With this bit-compliant model the designer can study all the effects that appear when a fixed word length is used. After converting the bit-compliant model into an executable model, it can be integrated into the same simulation scenario and checked against the user's requirements. The transformation of the functional- or bit-compliant model into an executable model requires code generators and hardware synthesis tools. We describe this in the next sections.

4.2. Software Generation

The code generation is based on the architecture shown in Fig. 5. The application software is based on a hardware independent layer which provides basic I/O and communication services, and a small real-time kernel. Thus, the generated C code can be easily ported to any node of the system. Third party tools are used to generate the executable code.

The actual implementation of the real-time kernel does not support true multitasking. Tasks are executed either as response to an interrupt or cyclically controlled by a timer.

4.3. Hardware Generation

We defined several steps to convert the bit-compliant model of a hardware task into the architecture-compliant model:

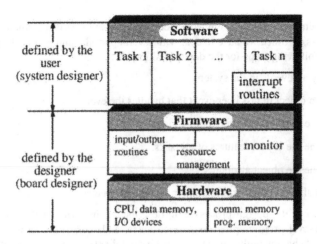

Figure 5. Programming model of a node processor.

1. Specification of the instruction set for the particular application based on a generic architecture.

2. Conversion of the bit-compliant model into an intermediate format (IAL = IMES Assembly Language) based on the instruction set definition.

3. Generation of the architecture based on the generic architecture, the instruction set definition, and the IAL-implementation of the data processing algorithm.

4. Mapping of the generated architecture onto an emulation system.

The generic processor architecture (Fig. 6) of the so called Application Specific Embedded Controller (ASEC) reflects the requirements of our domain of applications. It has four main building blocks: datapath, controller, memory, and process communication unit. The process communication unit comprises all functions necessary to read data from A/D-converters or digital I/O-ports into the internal memory or to write data to interface elements like D/A-converters or digital ports.

Fig. 7 shows an example of an instruction set definition. The first two lines define the multiplier type (LOC=I:1: Shift&Add multiplier). The third line specifies the adder type (LOC=CLA: Carry-Look-Ahead adder). The mnemonic of the instruction is given in brackets (e.g. [MACCS]: Multiply ACCumulate and Shift). The lines starting with S: address the arithmetic/logic elements necessary to execute the instruction. The lines starting with P: define the possible data sources and destinations for the instruction. The lines starting with V: specify the control flow during the execution of the instruction. After the designer has specified his specific instruction set, he has to convert the bit-compliant model into the IAL format.

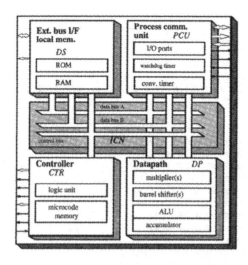

Figure 6. ASEC building blocks.

By using the program **IAL2XNF**, the architecture of an application specific processor is generated from the IAL description. The designer controls the transformation process simply by specifying design constraints like maximum internal bus width, number, and type of arithmetic blocks, etc. Different design alternatives can be evaluated quickly with regard to complexity and number of clock cycles. **IAL2XNF** analyses the data dependencies and utilizes the parallelism of the data processing algorithm in order to minimize the number of clock cycles.

4.3.1. Hardware Partitioning

Once the architectural description is generated, the design has to be partitioned in a way that it can be executed on the emulation board.

Most published partitioning approaches for FPGA based systems perform the partitioning after technology mapping, that is at the level of FPGA building blocks (e.g. Kuznar, Brglez and Kozminski, 1993). The main goal is to minimize the number of FPGAs and the number of interconnections under given timing constraints. Highly sophisticated heuristics are necessary to partition a netlist at the level of FPGA building blocks. This usually results in long execution times. Our domain of applications has some special characteristics that allow us to follow a totally different strategy:

- Rapid prototyping or ASIC emulation is based on a fixed emulation board architecture with a fixed number of FPGAs and a flexible number of interconnections between the FPGAs on the board.

```
MULOPT1:LOC=I:1
MULOPT2:LOC=I:1
ADDOPT1:LOC=CLA
[MACCS]
S:MUL1,MUX2,MUX3,MUX4,ADD1,REG1,BSH2,BUF3
P: A,BC,C,AB
B:2:3
V: 1: SYSRAM_A_ADR = $VAR1$;
V: 1: SYSRAM_B_ADR = $VAR2$;
V: 1: SYSRAM_A_WE = 0;
V: 1: SYSRAM_B_WE = 0;
V: 1: goto $2$;
V: 2: SYSRAM_A_ADR = $VAR1$;
V: 2: SYSRAM_B_ADR = $VAR2$;
V: 2: SYSRAM_A_WE = 1;
V: 2: SYSRAM_B_WE = 1;
V: 2: goto $3$;
V: 3: SYSRAM_A_ADR = $VAR1$;
V: 3: SYSRAM_B_ADR = $VAR2$;
V: 3: SYSRAM_A_WE = 0;
V: 3: goto $4$;
#INCLUDE
V: 4: IF SYSMUL_READY == 0 THEN $4$ ELSE $5$
V: 5: SYSRAM_A_ADR = $VAR1$;
V: 5: SYSRAM_B_ADR = $VAR2$;
V: 5: SYSRAM_A_WE = 1;
V: 5: SYSRAM_B_WE = 1;
V: 5: goto $NEXT$;

[BINT0]
S: COMP
P:L
V: 1: if (INT0 == 0) then $1$ else $LABEL$
```

Figure 7. Example of an instruction set definition.

- The architectures that are to be mapped onto the emulation board mainly consist of quite large arithmetic and/or logic building blocks and internal RAM/ROM. These blocks are connected to one or more internal buses.

This leads to the following partitioning strategy (Fig. 8):

1. Estimate the complexity of all entities in the netlist

2. Find clusters along buses with a complexity less than the complexity of a single FPGA.

3. Cut the nets between the clusters, and generate all necessary I/Os including their local control circuitry.

4. Place all clusters.

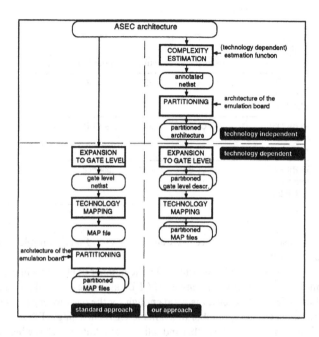

Figure 8. Partitioning approaches.

5. Generate the output files.

If the complexity of a single element (e.g. 32-by-32 parallel multiplier) exceeds the complexity of a single FPGA, then bipartitioning at CLB-level is performed after module expansion.

We developed two programs to support the described partitioning strategy: **CAP***estx*[7] and **CAP***mapx*[8]. **CAP***estx* estimates the number of CLBs for every entity in the netlists based on functions derived from actual implementations of macrocells with different bit-widths. For example, the complexity functions for parallel multipliers are as follows (see also Fig. 9):

$$
\begin{aligned}
fCLB\_est \quad &= 4.5491 \cdot n + 0.7917 &&\textit{for } m = 1\\
&= 0.983 \cdot n^2 + 1.5818 \cdot n - 4.1265 &&\textit{for } m = n\\
&= a + b \cdot n + c \cdot n^2 + d \cdot m + e \cdot n \cdot m + f \cdot n^2 \cdot m + g \cdot m^2 + h \cdot n \cdot m^2\\
&\quad m = \textit{external bit-width, } m = \textit{internal bit-width}
\end{aligned}
$$

$$a = 58, 86, \quad b = 6, 29, \quad c = 0, 11, \quad d = -21, 33,$$
$$e = 0, 80, \quad f = -0.02, \quad g = 1, 95, \quad h = 0, 01$$

$$
\begin{aligned}
f\,Reg\_est &= XE \cdot n + YE \cdot n + ar \cdot 2 \cdot n + 3, 2419 \cdot n + 8 &&\textit{for } m = 1\\
f\,Reg\_est &= XE \cdot n + YE \cdot n + ar \cdot 2 \cdot n &&\textit{for } n = m
\end{aligned}
$$

Since only simple functions have to be evaluated to calculate the number of CLBs the

65

Table 1. Comparison of CAPestx and PPR.

Example architecture	CAPestx		PPR (Xilinx)	
	Estimation	Runtime	Estimation	Runtime
8 Bit μC-architecture (clutch control)	149 CLBs	2' 33"	141 CLBs	4' 15"
16 Bit DSP architecture (digital controller)	273 CLBs	1' 59"	269 CLBs	11' 22"
32 Bit DSP architecture (IIR-Filter)	676 CLBs	1' 44"	abborted	7' 36"

execution is much faster than the (more accurate) estimation function implemented in the Xilinx PPR tool (see Table 1).

In **CAP***estx* the runtime is mainly determined by the number of elements in the design. In all our test applications the estimated number of CLBs always exceeds the real number of CLBs by less than 10%. This accuracy is good enough for the partitioning process because the user can also adjust the max. number of available CLBs in a single FPGA to guarantee routability. **CAP***mapx* partitions the netlist and introduces additional I/Os where necessary. The user can also preplace certain blocks to speed up the partitioning process.

After finishing partitioning several netlist files have to be expanded to gate level and mapped onto the internal structure of the FPGAs.

4.3.2. Module Expansion

The refinement of an architectural description to gate level is not very well supported in most FPGA design systems. Most of the things have to be carried out by hand. Module generators can help to automate this step. In the frame of our project a module generator (**CAP***blox*[9]) for XILINX® FPGAs was developed. The actual implementation supports 44 parametrisable macro cells:

- Input/output cells (INPUT, OUTPUT, OUTPUT-T, IO, INPORT, OUTPORT),

- select cells (TBUF, MUX, MAXU, MAXS, MINU, MINS),

- comparators (COMP, CLTGT, CLACBE, CLACAE, CLACLE, CLACGE),

- basic logic functions (AND, OR, XOR, NAND, NOR, XNOR, NOT, INV),

- basic arithmetic functions (ADD/SUB (riple-carry), CLAADD/ADC/SUB/SBB (Carry-Look-Ahead), MUL, BRLSHIFT, CONST),

- memories (RAM, ROM, REGISTER, SHIFTREG),

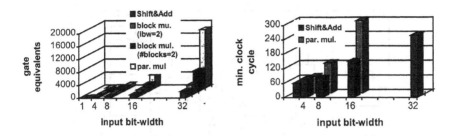

Figure 9. Cost functions for different types of multipliers.

- inference cells for hardware implementations of fuzzy logic (FBIU, FBIS),

- special cells (CONTROLLER) to include HDL descriptions (e.g. ABEL-HDL).

It should be noted here that **CAP***blox* includes generators for different types of multipliers and for fuzzy logic cells. Both are not supported in the actual version of XBLOX. Fig. 9 shows the complexity and cycle times of different multipliers generated by the **CAP***blox* multiplier generator. It shows that the complexity and cycle time of a shift-and-add multiplier grows linear with the input bit-width while the parallel multiplier shows a quadratic behavior.

CAP*blox* accepts XILINX® netlist format (XNF) as input format, which may be either generated from a graphical description or from synthesis tools (e.g. **IAL2XNF**). When module expansion is finished Xilinx tools can be used to perform the technology mapping and to generate the FPGA configuration bit stream.

4.3.3. ASEC Emulation

The emulation board (Herpel et al., 1993) is embedded in a heterogeneous multiprocessor environment which provides the interface to the development computer (PC-AT). This interface is used to download the configuration file to the board, to transfer data from the development computer, and to read back results from the emulation board. Therefore, the architecture-compliant model can be integrated into the simulation scenarios developed for the functional- and bit-compliant models, and checked against the user's requirements. The emulation speed is much faster than any simulation but usually slower than the real chip. For many applications, where the process time is slow (in the order of ms) the architectural model is equivalent to the timing-compliant model (not internally, but at the interface between ASEC and its environment!).

The emulation board itself consists of three main building blocks (Fig. 10): *logic emulation unit, process communication unit,* and *memory unit*. The memory unit provides buffers for the communication with other processors and local memory (up to 256 Kbytes) that can be used either as microcode or data memory. Different types of communication buffers (FIFO, dual port RAM, direct path), and synchronization mechanisms (mailbox,

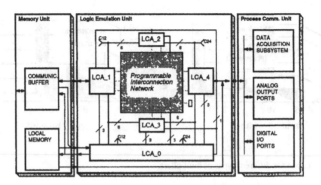

Figure 10. Emulation Board Architecture

Figure 11. Logic emulation unit.

semaphore) are integrated on a single chip (**IMES**-*ICM*) to support different communication mechanisms. The process communication unit supports the communication with external sensors and actuators. The logic emulation unit (Fig. 11) mainly consists of five FPGAs (XC4005) and four programmable interconnection chips (**IMES**-*CP5*16). The combination of fixed and programmable interconnections between the FPGAs allows for a high utilization and, thus, a net gate capacity of approx. 20 k gates.

5. Examples

5.1. Adaptive Pulsation Reduction in Hydraulic Systems

One goal of the mechatronics project "*Active pulsation absorber for hydraulic systems*" is to develop a highly integrated system for on-line computation and control of the volume

Table 2. Complexity and execution times of different design alternatives.

No.	Design alternative	Gate complexity	# cycles	Execution time/ emulation clock
1	1 parallel-multiplier	532 CLBs	50	25 μs / 2 MHz
2	1 block multiplier (2 blocks)	468 CLBs	90	45 μs / 2 MHz
3	1 Add&Shift multiplier	322 CLBs	210	42 μs / 5 MHz
4	TMS320C25		80	8,0 μs / 20 MHz
5	TMS320C30		80	4,8 μs / 33 MHz

flow in hydraulic systems (Herpel, Windirsch, and Glesner, 1992). After simulation at functional and bit-compliant level an assembly language implementation of the measurement and control algorithm was used to generate an application specific processor architecture. Simply by specifying constraints like maximum internal bus width, or type of multiplier different architecture variants could be generated within minutes. From the same high-level description, code for different DSPs was generated and analyzed for comparison. Thus, an early trade-off between speed and area was possible (Table 2).

In this particular case, a single chip solution with various interfaces (analog and digital) was required. The solution with a single add&shift multiplier fulfills the timing requirements (< 100 μs) at the lowest possible complexity. We mapped this architecture onto the emulation board and performed intensive testing before the actual chip was built.

The transformation of the emulated architecture into a standard cell design with the ES2 1.5 μm CMOS library had to be carried out by hand. At this point the design cycle is broken, but a shift from netlist formats to VHDL will allow to automate this design step in the future.

5.2. Self-Supervising Clutch

Friction clutches are machine parts used for connecting and disconnecting flows of torque between driving and driven shafts. A simple drive system generally consists of a motor, a clutch, one or more rotating masses, and a load. The engagement of the clutch is initiated from outside the system. During engagement of the clutch there is a transient slip, that causes heat in the contact areas. Wear in these areas increases dramatically with high temperature. Shortened life or even total failure are the results. By means of altering the temporal level of torque (torque characteristic) during the engagement phase the temperature can be reduced by 10 to 30 percent.

The overall purpose of the research project is to create an *intelligent* or *self-supervising* friction clutch. It is to control the engagement phase by selecting a proper temporal torque level considering the following points:

- The clutch must not heat up more than allowed.

- The impact of torque in the system is to be limited.

- Built-in capabilities for measuring and computing are to be used for torque-sensitive supervision similar as in a slip clutch or break-bolt clutch (automatic clutches).

A prototype of a clutch and a test rig were constructed, manufactured, and then successfully tested for mechanical function. The self-supervising clutch is a typical mechatronic system with mechanical, computational and cognitive processes.

The test rig is equipped with sensors for torque and speed of rotation in both driving and driven shaft, as well as three sensors for friction temperature at different points of the contact area. Thus, seven analog sensor signals can be obtained and digitized for further processing. The subsystem for influencing the torque during engagement is formed by quick piezoelectric actuators. Additionally, the initiation of the engagement as well as complete disengagement is done by a pneumatic cylinder. This is necessary because of the small displacements possible to be managed with piezoelectric actuators.

The model of the information processing unit consists of three concurrent tasks. The task "Perception" is assigned to read the sensor signals and build an internal representation of the clutch status. Sixty-four states can be distinguished in total. The engagement phase can last for about 0.3 to 10 seconds. Based on this internal model, user inputs and programmed behavior the task "Cognition" plans all necessary actions. This includes the calculation of an appropriate torque level to minimize the surface temperature and decisions about disengagement. The third task "Execution" controls the piezoelectric actuators. A rate of at least 5 kHz for reading sensor signals, computing, and writing actuator signals is required.

The behavior of the clutch was well understood and engineers from the mechanical engineering department developed a control algorithm. By talking to these engineers we found that fuzzy logic could solve at least one of their problems where they put a lot of effort in: temperature reduction by optimizing the shape of the temporary torque level. We realized and compared both, the approach with a conventional control algorithm and the approach based on fuzzy logic.

5.2.1. The "Conventional" Solution

If all the knowledge about the technical process, e.g. actual load, surface characteristics, actuator characteristics, etc. is available a priori, then a conventional solution, i.e. programming a microcontroller for example, can be realized in very short time. Depending on requirements such as costs, safety, reliability, physical size, etc. either a processor based or ASIC based solution can be selected.

In this particular case we had strong requirements concerning physical dimensions, number of components and reliability because it was planned to integrate the electronic into the

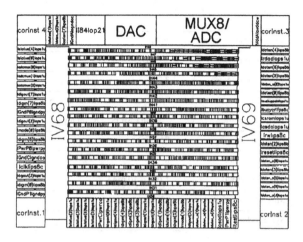

Figure 12. ASIC floorplan.

clutch. Based on these requirements, it was decided to implement the control algorithm as application specific processor. The design of this "embedded" controller was carried out in two phases: a prototyping phase and an ASIC synthesis phase. The main goal of the prototyping phase was to validate user requirements under real-time conditions.

Before any real-time code was generated, a functional compliant model was constructed and simulated under **CAP***view*. Since no complete clutch model exists, no closed-loop simulation was possible. The simulation was used only to ensure correct responses of the computational model to various input conditions (Fig. 13(a)). In the next step the functional model was refined to a bit-compliant model and executed on a DSP-board of **CAP***rTex*. The DSP-board was connected to the clutch testbed. Thus, we performed a hardware-in-the-loop simulation with a sampling frequency of 2 kHz. Fig. 13(c) shows the torque level and the resulting rotational speed of the driving and the driven shaft taken from this experiment.

For the ASIC implementation it was decided to use CALLAS, a high level synthesis tool from Siemens. CALLAS transforms a behavioral description into a set of gate primitives. Since the interface between the **CAP***code* code generator and the CALLAS synthesis system was not available at that time, manual transformation of the system description into DSDL code was performed. Due to the very similar language structures, this was mainly a search-and-replace operation with a text editor. As a result for the algorithmic circuit description, 430 lines of CALLAS DSDL code were generated and synthesized into a digital core cell with 1,648 standard cells.

By using the SOLO2030 design tool analog components (8-channel A/D converter and D/A converter) were added to the synthesized design and, thus, a 2-chip solution becomes possible: the controller chip itself and a ROM holding the temporal torque level, calculated beforehand from given clutch parameters. Fig. 12 shows the floorplan of the ASEC.

71

Table 3. Chip characteristics.

Chip characteristics	Value
# transistors	16.196
# gate equivalents	3.799
Chip size	5,629 mm * 4,237 mm = 23,85 mm^2
Technology	1,5 μm CMOS, 2 metal layer
Package	84 LCC
# analog inputs	8
# analog outputs	1 (max. 10 mA)
Max. data acquisition rate	15 kHz/channel
System clock	1 MHz

Table 3 gives some statistical information about the chip containing the synthesized digital core cell as well as the analog interface components.

5.2.2. *The Fuzzy Logic Approach*

A rule based solution does not require that all process parameters are available a priori, instead expert knowledge has to be formulated as a set of rules. The major advantage of a rule based system is, that such kind of systems are able to adapt themselves to changing parameters. One major drawback is that no formal methods exist to proof the safety of the system.

The application of fuzzy logic theory in control applications starts with the formulation of linguistic rules. These rules determine the behavior of the control unit, e.g. "If temperature is high, then torque is low". In a next step the designer has to specify, what is meant by the linguistic variables like "high", i.e. the corresponding graphs of the membership functions $\mu_I(x_j)$ are defined. This procedure leads to a description of the operator's behavior.

In total, the rule-base for the self-supervising clutch consists of 18 membership functions and 15 rules. Results from a simulation with this rule-base are shown in Fig. 13(b).

The results that were obtained during simulation show that the fuzzy logic approach is comparable to the conventional algorithm. The development of the rule-base for this application took about one week. This was about the same time we needed to optimize the shape of the temporary torque level. It should be noted here that the rule-base remains invariable while the parameters for the optimal temporal torque level have to be adapted whenever there is a change in the process parameters. More details about the fuzzy approach can be obtained from (Herpel et al., 1994).

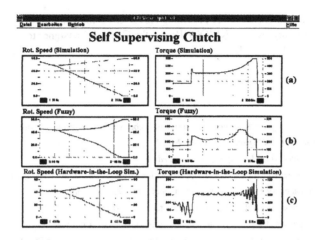

Figure 13. Simulated and measured signals.

6. Conclusions

MCEMS was used in several mechatronic projects like on-line process identification for combustion engines (Herpel, Windirsch and Glesner, 1991), active pulsation absorber (Herpel, Windirsch and Glesner, 1992), self-supervising clutch (Herpel et al., 1993), tire with integrated sensor for friction monitoring (Halgamuge, Herpel and Glesner, 1993), high precision measurement instrument, and also in microsystems technology (Herpel et al., 1995).

All these projects showed that it was very helpful to have a prototype during early design phases. System engineers could play with the system and try to find a near optimal solution. The lengthened specification phase led to real-time information processing units which need no further modifications when integrated into their environment. Thus, the overall development cycle was shortened.

The models we use seem to be appropriate for our domain of applications, but the supporting tools need to be improved. Future work will concentrate on the development of synthesis tools supporting VHDL rather than netlist formats. This will make the transition from prototype to product much easier. Additional work is necessary to support the partitioning of a system into concurrent tasks.

Acknowledgments

The authors like to thank the DFG for sponsoring the project and all collaborators in the project (Prof. Dr.-Ing. B. Breuer, Dr.-Ing. J. Stöcker; Prof. Dr.-Ing. Dr. h.c. mult. R. Isermann, Dr.-Ing. J. Bußhardt, Dr.-Ing. J. Führer; Prof. Dr.-Ing. Dr. h.c. mult. G. Pahl, Dr.-Ing. W. Habedank; Prof. Dr.-Ing. B. Stoffel, Dr.-Ing. K.J. Kurr) for their kind cooperation.

They also want to thank the EU for providing ASIC- CAD tools in the frame of COMETT
VLSI Design Training Action EUROCHIP and the CALLAS designer team at SIEMENS
AG, Munich for their support.

Notes

1. MCEMS = Methodology for the Conceptual Design of Embedded Microelectronic Systems
2. DIN= german acronym for "Deutsche Industrie Norm"
3. CAP*edit* = Computer Aided Prototyping graphical *editor*
4. CAP*view* = Computer Aided Prototyping support tool for *virtual process reality under windows*
5. CAP*rTex* = Computer Aided Prototyping support tool for *real-Time execution*
6. CAP*code* = Computer Aided Prototyping *code* generation environment
7. CAP*estx* = Computer Aided Prototyping *estimation tool for xilinx netlist format*
8. CAP*mapx* = Computer Aided Prototyping *mapping* of netlists onto multiple *xilinx* FPGAs
9. CAP*blox* = Computer Aided Prototyping by *block synthesis to xilinx netlist format*

References

Adé, M., Wauters, P., Lauwereins, R., Engels, M., and Peperstraete, J. A. 1993. Hardware-software trade-offs for
   emulation. L. D. J. Eggermont, P. Dewilde, E. Deprettere, J. van Meerbergen (eds.), *VLSI Signal Processing*,
   IV. New York: IEEE.
Boehm, B. W., Gray, T. E., and Seewaldt, T. 1984. Prototyping versus specifying: A multiproject experiment.
   IEEE Transactions on Software Engineering SE-10(3): 290–302.
Färber, G. 1992. *Prozeßrechentechnik*. 2. Aufl. Berlin: Springer-Verlag.
Gomaa, H. 1990. The impact of prototyping on software system engineering. R. H. Tayer, M. Dorfman (eds.),
   System and Software Requirements Engineering, Washington D.C.: IEEE Computer Society Press, pp. 543–552.
Gordon, V. S., and Bieman, J. M. 1995. Rapid prototyping: Lessons learned. *IEEE Software* Jan., pp. 85–95.
Halgamuge, S. K., Herpel, H.-J., and Glesner, M. 1993. An automotive application based on extracted knowledge.
   Proc. of the International Workshop on Mechatronical Computer Systems for Perception and Action Halmstadt,
   June, pp. 295–299.
Herpel, H.-J., Glesner, M., SÜß, W., Gorges-Schleuter, M., and Jakob, W. 1995. Rapid prototyping in microsys-
   tems development. Accepted for publication on the *6th IEEE Workshop on Rapid System Prototyping*, Chapel
   Hill, June.
Herpel, H.-J., Halgamuge, S. K., and Glesner, M., Stöcker, J., and Ernesti, S. 1994. Fuzzy logic applied to
   data analysis problems in automotive applications. *Proc. Mechatronics & Supercomputing Applications in the
   Transportation Industries of the ISATA'94* Aachen, Nov., pp. 225–232.
Herpel, H.-J., Held, M., and Glesner, M. 1994. MCEMS toolbox—A hardware-in-the-loop simulation environment
   for mechatronic systems. *Proc. IEEE International Workshop on Modeling, Analysis and Simulation of Computer
   and Telecommunication Systems* Durham, Jan., pp. 356–357.
Herpel, H.-J., Held, M., and Glesner, M. 1993. Real-time system prototyping based on a heterogeneous multi-
   processor environment. *Proc. 5th Euromicro Workshop on Real Time Systems* Durham, June, pp. 62–67.
Herpel, H.-J., Wehn, N., and Glesner, M. 1995. Computer aided prototyping of application specific embedded controllers in mechatronic systems. J. Rozenblit, K. Buchenrieder (Hrsg.), *Codesign: Computer Aided
   Software/Hardware Engineering*, New York: IEEE Press, pp. 425–442.
Herpel, H.-J., Wehn, N., Gasteier, M., and Glesner, M. 1993. A reconfigurable computer for embedded control
   applications. *Proc. IEEE Workshop on FPGAs for Custom Computing Machines* Napa, April, pp. 111–120.
Herpel, H.-J., Windirsch, P., and Glesner, M. 1992. ASIC based volume flow measurement in a mechatronic
   system. *Proc. 3rd EUROCHIP Workshop on VLSI design training* Grenoble, Sept., p. 290.

Herpel, H.-J., Windirsch, P., and Glesner, M. 1991. A VLSI implementation of a state variable filter algorithm. *Proc. IEEE Symposium on VLSI* Kalamzoo, March, pp. 138–143.

Herpel, H.-J., Windirsch, P., and Glesner, M. Habedank, W., and Pahl, G. 1993. Self supervising clutch— An integrated mechano-electronic solution. *Proc. of the International Workshop on Mechatronical Computer Systems for Perception and Action* Halmstadt, June, pp. 339–346.

Hsia, P., Davis, A., and Kung, D. 1993. Status report: Requirements engineering. *IEEE Software* 10(6).

Kuznar, R., Brglez, F., and Kozminski, K. 1993. Cost minimization of partitions into multiple devices. *Proc. of the 30th Design Automation Conference,* Dallas.

Lugi, and Ketabchi, M. 1988. A computer aided prototyping system. *IEEE Software* March, pp. 66–72.

Walters, S. 1991. Computer-aided prototyping for ASIC-based systems. *IEEE Design & Test of Computers* June, pp. 4–10.

Real-Time Systems, 14, 293–310 (1998)

Implementation of Hard Real-Time Embedded Control Systems *

MATJAŽ COLNARIČ   colnaric@uni-mb.si

DOMEN VERBER

ROMAN GUMZEJ
Faculty of Electrical Engineering and Computer Science, University of Maribor, 2000 Maribor, Slovenia

WOLFGANG A HALANG   wolfgang.halang@fernuni-hagen.de
Faculty of Electrical Engineering, FernUniversität, 58084 Hagen, Germany

Abstract. Although the domain of hard real-time systems has been thoroughly elaborated in the academic sphere, embedded computer control systems — being an important component in mechatronic designs — are seldom dealt with consistently. Often, off-the-shelf computer systems are used, with no guarantee that they will be able to meet the requirements specified. In this paper, a design for embedded control systems is presented. Particularly, the paper deals with the hardware architecture and design details, the operating system, and high-level real-time language support. It is shown how estimates of process run-times necessary for schedulability analysis can be acquired on the basis of deterministic behavior of the hardware platform.

Keywords: embedded computer control systems, hard real-time systems, microcontrollers, transputers, earliest-deadline-first scheduling, real-time programming languages

1. Introduction

Already being thoroughly elaborated in the academic sphere, consistent hard real-time systems design is now slowly making its way towards the applications. Per definition, embedded hard real-time systems are employed to control a variety of technical applications; they represent one component of mechatronic systems' co-design, together with the mechanical and electrical parts.

Often the integrity of such systems is safety-critical: a failure may result in major material loss, or even endangerment of human lives. While in these systems testing of conformance to functional specifications is well established, temporal circumstances, being — at least — an equally important part of a design, are seldom consistently verified. It is almost never proven at design time that such a system will meet its temporal requirements in every situation that it may encounter.

Seemingly the most characteristic misconception in the domain of hard real-time systems (Stankovic, 1988) is that real-time computing is often considered as fast computing. It is obvious that computer speed itself cannot guarantee that specified timing requirements will be met. Instead, a different ultimate objective was set: predictability of temporal behavior. Being able to assure that a process will be serviced within a predefined time frame is of utmost importance. In multiprogramming environments this condition can be expressed

---

* Work on this project has been partly supported by the Ministry of Science and Technology of the Republic of Slovenia

as schedulability: the ability to find, a priori, a schedule such that each task will meet its deadline (Stoyenko, 1987).

For schedulability analysis, execution times of hard real-time tasks, whose deadlines must be met in order to preserve the integrity of a system, must be known in advance. These, however, can only be determined if the system functions in a deterministic and predictable way in the temporal domain. To assure overall predictability, all system layers must behave predictably in the temporal sense, from processor to system architecture, operating system, language, and exception handling (layer-by-layer predictability, (Stankovic, 1990)). In addition to hard real-time tasks, soft real-time tasks may be run in idle times. Their integrity is less critical and deadlines may sometimes be missed without major consequences; for them, top-layer predictability (Stankovic, 1990) is achievable. This paper, however, is only dealing with the critical parts which must behave deterministically, and whose temporal behavior must be easily and not too pessimistically predictable.

The results of fundamental research in this domain are still not broadly used in hard real-time applications. One of the reasons is that most of the studies only consider selected topics and assume that the rest behaves predictably. This assumption makes it very hard for application designers to set up a consistent system from different conceptions.

Following the ultimate objective of mechatronic design, we decided to initiate a project where all crucial layers of hardware and software components are systematically addressed in a holistic manner to provide practically usable hard real-time control system design techniques. Our objective was not to generate novel approaches or methods, but to select and consistently use concepts from our and other groups' previous research.

In this sense we decided to use off-the-shelf processors and other hardware components; they were carefully selected according to their temporal determinism. To achieve temporal predictability and to optimise performance, hardware and software were co-designed in the large and in the small scale: their features mutually support each other.

2. Prototype of Hardware Platform

In multitasking process control systems, dynamic scheduling algorithms are required. Therefore, we did not make use of rate monotonic scheduling, which is receiving much attention in the literature. For our purpose, among the feasible strategies the earliest-deadline-first algorithm was chosen and implemented in the kernel of an operating system (Colnarič, Halang and Tol, 1994). It has been shown (Henn, 1989) that it provides the highest processor utilization and the lowest number of pre-emptions among all task scheduling disciplines feasible for single processor systems, and that it is consistent with a straightforward necessary and sufficient schedulability analysis criterion holding on the task level; with the throw-forward extension it is also feasible on homogeneous multiprocessor systems. However, this more complex extension leads to more pre-emptions and is, thus, less practical.

For process control applications, where process interfaces are usually physically hardwired to sensors and actuators establishing the connection to the environment, it is natural to implement either single processor systems or dedicated multiprocessors acting and being programmed as separate units. Thus, the earliest-deadline-first scheduling policy can be employed without causing any restrictions, resulting in a number of advantages as discussed by Halang and Stoyenko (1991).

In the classical computer architecture the operating system is running on the same processor(s) as the application software. In response to any occurring event, the context is switched, system services are performed, and scheduling decisions are made. Although it is very likely that the same process will be resumed, temporal determinism of process execution is violated and a large amount of performance is wasted by superfluous overhead. This suggests employing a second, parallel processor to carry out the operating system services. Such an asymmetrical architecture turns out to be advantageous, since, by dedicating a special-purpose, multilayer processor to the real-time operating system kernel, the user task processor(s) is (are) relieved from any interruptions and administrative overhead.

This concept was in detail elaborated in (Halang, 1988; Colnarič, Halang and Tol, 1994). It was also independently considered in a number of research projects and implemented in several successful prototypes (e.g., (Lindh, 1989; Roos, 1990; Stankovic, 1995; Cooling, 1993)). Compared to (Cooling, 1993), it is more complex and supports multiprocessing and, to a certain extent, also distributed processing. However, it is still targeted to embedded applications and is, thus, less sophisticated than (Stankovic, 1995). The latter supports local area networking and features flow control filters, reflective memories, and a VLSI implemented scheduling controller.

The experimental hardware platform described in this paper complies to a high degree with the principles described in (Halang, 1988; Colnarič, 1994). In Figure 1 it is shown that it consists of task processors (TPs) and a kernel processor (KP) which are fully separated from each other. Loose coupling is achieved by serial links and few additional signals; this promises easy local distribution of process control (task) processors, and migration of processing power close to where it is needed.

The **kernel processor** (KP) is responsible for all operating system services. It maintains the real-time clock, and observes and handles all events related to it, to the external signals and to the accesses to the common variables and synchronizers, each of these conditions invoking assigned tasks into the ready state. It performs earliest-deadline-first scheduling on them and offers any other necessary system services.

The external process is controlled by tasks running in the **task processors** (TP) without being interrupted by the operating system functions. A running task is only pre-empted if, after re-scheduling caused by newly invoked tasks as a consequence of an event, it is absolutely necessary to assign the highest priority to it in order to ensure that all tasks will meet their deadlines.

Although the real-time clock is really necessary only in the kernel processor (and optionally in peripheral controllers to allow for exactly timed and jitter-free I/O operations as described in (Halang, 1989)), it may be available as relative Julian time throughout the system: in any of the interested units it can be implemented in hardware or in software as a counter of standard time unit ticks ($RTclock$ signal), generated by the kernel processor and beginning at system initialization. Absolute time can easily be calculated from the counter value and the absolute time recorded for the system initialization instant.

In process control it is often necessary to perform certain operations at exactly defined instants. As a rule, this cannot be achieved with high accuracy by means of scheduling tasks to be activated at those times because of operating system overhead and non-deterministic

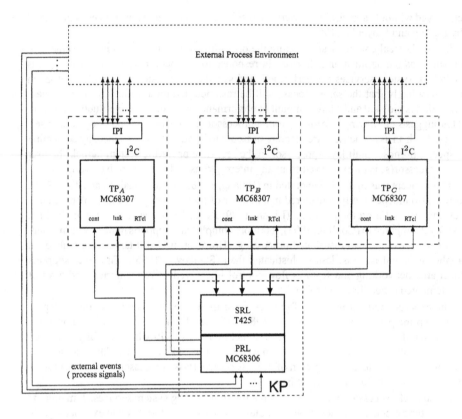

Figure 1. General Model of the Hardware Architecture

precedence relations holding between the active tasks at those very moments — particularly not in dynamic environments. This option is implemented by halting task processing in a certain state and re-activating it by continuation signals generated by the kernel processors.

This concept is preventing non-deterministic interruptions from the environment to reach the task processors by (1) careful avoidance of the sources of unpredictable processor and system behavior, (2) by loose coupling of task processors, and (3) by synchronous operation of the kernel processor. The predictability of temporal system behavior achieved through this concept is expected to be sufficient to provide the necessary basis for the higher system design levels.

In the following sections the implementation of the hardware units will be detailed.

2.1. *Operating System Kernel Processor*

To support determinism and for better performance the operating system kernel processor is divided into two layers, a higher-level part called *secondary reaction layer (SRL)*, and

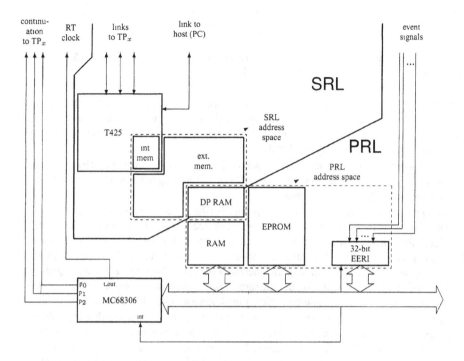

Figure 2 Operating System Kernel Processor

a lower-level part or *primary reaction layer (PRL)*, see Figure 2. Since their functions are tightly coupled, they communicate via a double-port memory $(DPRAM)$, where their interfaces reside in form of operating system data structures and the real-time clock storage element. Conflicts of simultaneous access can be resolved by an arbiter built inside the dual-port memory. However, accesses are exclusively performed in carefully defined time slots; this way conflicts are avoided.

Briefly, the PRL is performing elementary hardware functions like external event recognition, time and time event management etc., while the SRL serves as interface to the task processors, manages their operating system service requests and carries out the actual scheduling on the basis of information on events provided by the PRL.

There are several reasons for structuring the operating system processor into two parts. An important one is the natural parallelism between the lower-level (signal sensing, administration of real-time clock, time schedules etc.) and the higher-level functions (scheduling, operating system routines); exploiting that, the performance of the kernel processor is improved and, hence, the response times are shortened. Further, the primary reaction layer is intended to be implemented in an FPGA in the forthcoming versions of the prototype. Finally, in our particular implementation, the transputer provides no reasonable interrupt handling and its two level scheduling was found inappropriate for this purpose; thus, we decided to add the lower-level processor.

In the sequel, functions and implementation of the PRL and the SRL will be detailed.

2.1.1.   Primary Reaction Layer   The PRL is implemented using a Motorola MC68306 microcontroller. The basic functions of the **PRL** are:

• *Generation of the RTclock signal:* RT ticks are generated by the MC68306 built-in timer/counter. For every tick a preset value is loaded into the counter and decremented with every clock cycle; on its expiration, the *RTclock* tick is generated at the timer's external output and the internal interrupt is signaled.

• *Administration of the real-time clock in form of Julian time.* It is maintained by software: on every *RTclock* tick the time is incremented. The 32 bits long word allows relative times of 49 days with a resolution of 1 msec.

During power-on, all real-time counters throughout the system are reset and the *RTclock* signal is disabled. In the initialization sequence the kernel processor acquires precise actual absolute time (either from a host computer or via some other means — e.g., the Global Positioning System), adds the necessary (constant and known) overhead until the counters are started, and distributes it among all interested parties. Then, the *RTclock* signal is enabled and all counters begin to count the same ticks. Hence, the relative times will be the same, and the constant offset to the absolute time known all over the system.

• *Administration of time events introduced by time-related schedules* (either for tasking operations or for continuation of their execution): SRL provides a vector of activities which are scheduled to be activated at certain times. PRL administers this vector, selects the closest time event and, upon its occurrence, accordingly provides information so that the associated actions can be taken over by SRL.

• *Reception and handling of signals from the process environment (and monitoring whether overruns occur):* events from the environment are received by a specially designed external event recognition interface implemented by a XILINX LCA. It provides 32 parallel digital I/O ports; active (negative) transition on each of these ports sets a bit in a register which is periodically polled; any bits set are cleared and registered in the operating system data structures together with time stamps. If an event appears more than once during the same polling cycle, an overrun bit is set indicating that at least one event was not serviced.

• *Provision for exact timing and/or synchronous operation of task processors:* the already mentioned exact timing feature offers the possibility to schedule *a part of a task* to be executed at an exactly defined instant which, in principle, cannot be done by usual task-level scheduling on a time event. To achieve that, a task running in a task processor can halt its own execution using an operating system call. In the call a time schedule is associated with this action; when it is fulfilled, the *continuation* signal re-enables the program in the task processor by releasing the wait state. The continuation signals for all task processors are driven by the microcontroller's I/O lines and can be triggered individually or simultaneously by disjunctive combination of the I/O port data register.

• *Providing information on the actual events in the current system cycle to SRL* by putting them into the common data structures, to be available for actual scheduling of the associated tasks.

2.1.2. Secondary Reaction Layer   The SRL communicates with the task processors from which it receives calls for operating system service. These calls may require certain information about the state of the system or of tasks. Tasks may access the synchronizers and the common global variables residing in the SRL; a change in status of these structures triggers an event to which a task activation may be associated. Important types of service calls are tasking requests; a task running in a task processor may request time-related and/or event-related scheduling of different tasking operations. The conditions are passed to the PRL which triggers events upon their fulfillment.

When such events occur, the SRL invokes the tasks assigned to them, performs schedulability analysis, and actually places the ready tasks into the processor queue according to their deadlines.

The SRL is, from the viewpoint of the hardware designer, a standard transputer system with an external bus access to the dual-port RAM. It has three bi-directional links to task processors and one to a host for purposes of monitoring and program down-loading. The latter can also be used as a link to a fourth task processor once an application is operational and its programs were placed in ROMs. These serial point-to-point communication paths were, apart from adequate processing power, the main motivation for using transputers.

Although not very practical, the system can also be scaled to slightly larger applications by adding an additional transputer to one link of the secondary reaction layer. Thus, six task processors can be supported and further two with each additional transputer; however, this would influence the response times of the system due to the additional load. To this point, a number of three or four task processors supported in the basic configuration was found to be most reasonable with regard to overall system performance.

2.2. Application Task Processors

The actual process control programs are executed in task processors. Only a small part of the operating system resides here performing communication with the SRL and context-switching upon the latter's request. In a task processor's memory the program code of each task mapped to it is kept. Also, parts of these tasks' control blocks are residing there, holding the contexts of eventually pre-empted tasks.

Basically, a task processor can be implemented using a wide variety of processing devices, such as microcontrollers, digital signal processors, or automata based on FPGAs or special-purpose VLSI components. Important selection criteria are the envisaged purpose in the application domain, and determinism and predictability of the temporal program execution behavior. To support run-time analysis for such a diversity of processors, we introduce a configurable compiler with a built-in run-time estimator using different methods and techniques which will briefly be outlined in Section 4.

In Figure 3 the task processor architecture is shown as implemented in our prototype with the Motorola MC68307 microcontroller. The temporal execution behavior of the MC68000 microprocessor kernel is not excessively complicated; analyses have shown that it is predictable with sufficient ease.

Every message received on the transputer link interrupts task program execution. It can either be a call for pre-emption of the running task, or a response to an operating system

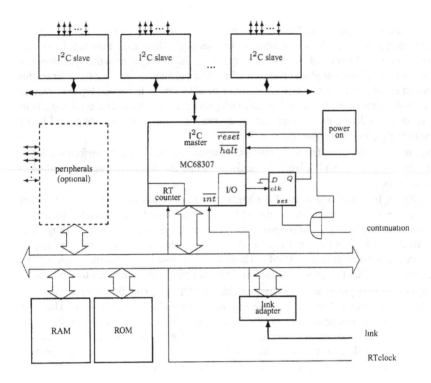

continuation

Figure 3. Application Task Processor

service request. In both cases the delay in task execution caused can be considered in the context-switch time, or in the operating system response time, respectively.

The microcontroller's internal timer can be used as a counter of *RT clock* ticks. Although the task processor conceptually does not need a real-time clock, its presence may save service calls to the kernel processor.

Below, some features of the task processor are presented.

• *Intelligent Process Interfaces* (IPIs) serve as interfaces between task processors and controlled environments. They are based on microcontroller or FPGA-supported peripheral devices connected to task processors via I^2C serial buses. Their services are available by calling pre-defined peripheral device drivers and providing parameters and data. IPIs provide higher-level peripheral interface services. By moving intelligence towards peripheral devices, task processors are relieved from that work, peripheral services are much more flexible and fault-tolerant, and system safety is enhanced.

Optionally, in simple implementations, instead of peripheral interfaces on an I^2C bus, peripheral devices may be directly implemented on task processor modules. For this purpose the MC68307 microcontroller's remaining peripheral interfaces may be used. Then, alternative peripheral device drivers must be prepared. On the other hand, in more demanding environments, peripheral interfaces may be linked to application task processors by sensor/actuator buses or, generally, field buses. Under the aspect of real-time capabilities, viz., bounded end-to-end communication delay and deterministically guaranteed access time to the transmission medium, the best performance among field buses is then offered by InterBus-S (Blume and Klinker, 1993).

Peripheral devices must be carefully selected and implemented in such a way that their timing behavior is deterministic and can be predicted. For instance, analoge-to-digital conversion can take different time in certain implementations depending on the input level. A consistent implementation of such a peripheral device is to permanently sample and convert the analogue value; upon request always the value acquired last is returned without any delay. Assuming that the sampling interval is sufficiently small, the result is acceptable.

• *Exact timing of operations:* this feature already mentioned in the discussion of the kernel processor requires a counter-part in the task processors. After a task, reaching a point from where continuation at a certain time instant is requested, has issued the corresponding service call to the kernel processor, it enters the wait state sending a *wait* signal to a simple circuit triggered by an I/O line. When the instant for resuming the task processor's operation and the execution of the task is reached, which is indicated by a time event, the PRL generates a *continuation* signal which resets the *wait* signal to the inactive state thus resuming operation.

While providing the most precise timing of process execution, e.g., for purposes of temporal synchronization, this implementation seems to have the drawback of halting the processor preventing other ready tasks from executing during the idle time until the moment of continuation. In practice, however, there is no blocking of other tasks, since the waiting times are kept short and are considered in the estimation of the corresponding tasks' run-times. As an alternative to this solution, and to prevent idle waiting, an exact timing feature is also implemented on the level of intelligent peripheral interfaces. Along with an I/O command, the time when the operation is to be performed is transmitted to an IPI, which has access to the real time clock. The calling task is suspended, waiting for the requested instant when the data is sent out or, respectively, latched into a buffer, ready to be read by the task processor. Which method is utilized depends on whether only peripheral operation or task execution as a whole is requested to be timed precisely.

With exact timing of input/output operations as introduced in (Halang, 1993) application designers can cope with the jitter problem: with a slight remaining jitter only depending on the real-time clock, they can precisely prescribe when certain sensors are read or actuators set in closed regulatory loops.

• *Distributed processing:* process control and embedded applications are often implemented as distributed control systems in rather natural ways. Our architecture matches this approach on two levels, viz., on the level of task processors and on the level of intelligent peripheral interfaces.

Task processors are connected to the kernel processor via transputer links which are limited in length; however, it is possible to implement them in fiber optics, thus allowing distribution over longer distances. The serial bus connecting peripheral interfaces with task processors works with lower transfer rates and is, thus, less problematic; commercial drivers allow transfer over reasonable distances. Since in both communication protocols only master-slave and single-master options are implemented, there is extremely low overhead in data transfer.

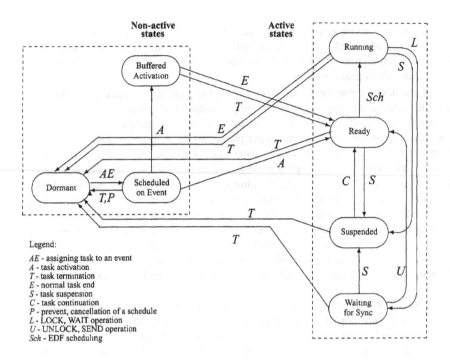

Figure 4. Task States and Their Transitions

3. Operating System Support

Basic and theoretical information about the operating system and its rationale was presented in (Colnarič, Halang and Tol, 1994). Here, we only mention that the earliest-deadline-first scheduling policy is implemented. To prevent unnecessary context-switches we introduced a simple modification: although a newly arrived task has an earlier deadline than the executing one, the running task is not pre-empted, unless the situation caused by the latest event is such that it requires immediate task re-scheduling. Apart from better performance due to considerably less pre-emptions, the problem of resource contention resulting from mutual exclusion of tasks using critical regions is minimized.

An application task, which exists inside, and is intended to be executed by a task processor, is in one of the following states (see Figure 4):

Dormant: Being in the dormant state means that a task's code exists, and the task control block (TCB) holds its parameters. Actually, the TCB is split into two parts, which hold (1) the system information about the tasks and (2), the tasks' current variable values, and reside in the SRL and in the task processor, respectively.

Scheduled on event: A task is assigned to an event; as soon as the latter occurs the task will be activated.

Ready: If a task's condition to be run is fulfilled, it is moved into the *ready* state. Each time a task enters this state a schedulability analysis is dynamically performed to check whether its deadline is closer than the one of the executing task and, further, whether its laxity allows the execution of the latter to be continued and, thus, a pre-emption to be avoided without violating any deadlines.

Buffered activation: Only one instance of a task may be active at the same time. If an already active task is to be activated anew due to fulfillment of some schedule, its activation is buffered, i.e., postponed until the already active instance is either completed or terminated.

Running: A task can only be put from the *ready* into the *running* state by the scheduler, which is part of the operating system being executed by the SRL.

Suspended: A task activation may be suspended. In this state a task retains its internal logical state; however, since suspension is not temporally bounded, timing circumstances become invalid and must be restored upon continuation of task execution by defining a new deadline.

Waiting for synchronization and critical regions: To synchronize with another one, a task may send, or wait for a signal. Also, to protect a part of a program, it may be enclosed into a critical region. If a certain signal is not present on "wait", or a task cannot enter an already occupied critical region, its state changes to *waiting for synchronization*.

Tasks change their states upon operations such as *activate, terminate, end, suspend, continue* (see Table 1) etc. These operations can be scheduled to be executed upon fulfillment of a condition, called event, or a combination thereof. Events can be time-related and, then, of sporadic or periodic nature; they may occur at certain instants or after specified time intervals. Non-time-related events are interrupts from the environment and value changes of shared variables. Time-related and non-time-related schedules may be combined. This way, temporally bounded waiting for events from the environment is enabled.

When a schedule is fulfilled, an associated task may be:

activated: the task is put into the ready state and, consequently, actually run;

terminated: the task is prematurely terminated and put into the dormant state; also, all schedules related to the task are cancelled;

suspended: the task's activation is suspended, but its internal logical state is preserved, temporal circumstances are disregarded and, thus, become invalid;

continued: after a suspension the task can be re-activated; a new deadline must be supplied;

resumed: in general, "resume" is a combination of "suspend" and "continue", i.e., the task is immediately suspended and is resumed when the associated schedule is fulfilled. The actual execution of this function is dependent on the option "exactly": if it is not selected, the task changes its states from "running" to "suspended" and again to "ready", as it would if the functions "suspend" and "continue" were called explicitly. However, when "exact" timing is requested, the task does not leave the "running" state: upon the function call execution is halted, and continued immediately after occurrence of the scheduling event, which may be time-related or not, without any delays for scheduling, context-switching and similar overhead. Waiting in the suspended state is temporally bounded by a "timeout" clause, being necessary to retain temporal predictability of task execution; if the timeout expires before the event has occurred, a given "ontimeout" sequence is executed. It is also possible to disregard temporal circumstances and set another deadline, which is reasonable when the "exactly" option is not chosen and the task must be re-scheduled for continuation.

Apart from the above, there are some other related actions, such as:

end: normal task end when execution is completed;
prevent: cancellation of certain schedules;
tstate, sstate, now: monadic operators to acquire task or semaphore states, or time;
enable, disable: external signal masking.

For consistent inter-task synchronization, send and wait functions are provided using signals. Further, mutual exclusion of tasks can be achieved using temporally bounded critical regions; they may be pre-emptable or not as indicated by an "np" option. If any non-pre-emptable critical region exists in a task set, the longest one must be considered in schedulability analysis (for details it is referred to (Colnarič, Halang and Tol, 1994)). Both features are implemented using semaphores and "lock" and "unlock" functions, which are also shown in the state transition diagram of Figure 4.

Both waiting for a signal and waiting to enter a critical region are temporally bounded by "timeout" clauses; corresponding "ontimeout" sequences need to be provided. Thus, synchronization delays become predictable and can be considered in task run-time estimation and further in schedulability analysis.

Similarly, it is possible to temporally bound any part of a program using a "during" construct. This has two functions: (1) to protect any temporally non-deterministic features (e.g., peripheral operations), and (2) to assert explicit estimation of a certain program part's run-time, whose automatically obtained estimate is expected to be imprecise and pessimistic. Automatic estimation of the run-time is overridden by this assertion.

4.   High-Level Real-Time Programming Language Support

As a part of our project, a new real-time programming language was developed and a compiler with a built-in run-time analyzer designed. Being aware of the unpopularity of producing new languages and writing own compilers, we considered the following arguments for doing so.

• In spite of the growing urge, we found that commercially available and widely used high-level programming languages and compilers still do not meet all the needs of embedded hard real-time application programming. Most languages used for such systems produce code that leads in one way or another to unpredictable timing behavior.

• Another pro-argument was the co-design of our particular experimental hardware architecture, the corresponding operating system and the application development tool. It is impossible to support the features of the former two with any existing compiler.

Scheduling support

```
<non_time_schedule > ::= WHEN <non_time_event> { OR <non_time_event> }
<non_time_event> ::= <interrupt_id> | <shared_variable> [<rel_op> <exp>]
<simple_time_schedule > ::= AT <time_exp> | AFTER <dur_exp>
```

```
<periodical_time_schedule>::=[<simple_time_schedule>]EVERY<dur_exp>[DURING<dur_exp>]
<time_schedule> ::= <simple_time_schedule > | <periodical_time_schedule>
<combined_schedule > ::= <non_time_schedule> OR <time_schedule>
<schedule> ::= <non_time_schedule> | <time_schedule> | <combined_schedule>
```

Task activation

```
<task_activation> ::= [<schedule>] <task_activation_statement>
<task_activation_statement> ::= ACTIVATE <task_id> DEADLINE IN <dur_exp>
```

Task termination

```
<task_termination> ::= [<schedule>] <task_termination_statement>
<task_termination_statement> ::= TERMINATE [<task_id>]
```

Task prevention

```
<task_prevention> ::= [<schedule>] <task_prevention_statement>
<task_prevention_statement> ::= PREVENT [<task_id>]
```

Task suspension

```
<task_suspension> ::= [<schedule>] <task_suspension_statement>
<task_suspension_statement> ::= SUSPEND [<task_id>]
```

Task continuation and resumption

```
<task_continuation> ::= [<schedule>] <task_continuation_statement>
<task_continuation_statement> ::= CONTINUE <task_id> DEADLINE IN <dur_exp>
<task_resume > ::= [EXACTLY] <schedule> <task_resume_statement>
<task_resume_statement> ::= RESUME [<task_id>] [DEADLINE IN <dur_exp>]
   TIMEOUT <dur_exp> [ONTIMEOUT <statement>]
```

Normal task end

```
<task_end> ::= <task_end_statement>
<task_end_statement> ::= END
```

Synchronization constructs and critical region concept

```
<send_statement> ::= SEND <signal_id>
<wait_statement> ::= WAIT <signal_id> TIMEOUT <dur_exp> [ONTIMEOUT <statement>]
ENTER <semaphore_id> [NP] TIMEOUT <dur_exp> <statement>{<statement>}
   [ONTIMEOUT <statement>{<statement>}] LEAVE
```

Explicit assertion of execution time

```
DURING <dur_exp> [NP] DO <statement>{<statement>}
   [ONTIMEOUT <statement>{<statement>}] FIN
```

Status or value acquisition constructs

```
<task_state_statement> ::= TSTATE <task_id>
<sync_state_statement> ::= SSTATE <sema_id>
<get_RTC_statement> ::= NOW
```

Interrupt enabling and disabling constructs

```
<interrupt_enable_statement> ::= ENABLE <interrupt_id>
<interrupt_disable_statement> ::= DISABLE <interrupt_id>
```

Table 1. Syntax of the Programming Language Features Supported by the Operating System

```
ADD16   MACRO
        MOVE.W   &1,D0
        ADD.W    &2,D0
        MOVE.W   D0,&1
        TIME 36
        ENDM
```

Figure 5. An Example for a Translation Macro

• Further, experience with run-time analysis of high-level source code programs yielded that access to the compiler-internal data structures is required not being possible for commercial compilers.

Owing to these considerations, we decided to design our own programming language miniPEARL based on the standardized programming language PEARL with certain modifications and simplifications (Verber and Colnarič, 1993; Verber and Colnarič, 1995). MiniPEARL is a strongly typed and structured high-level real-time programming language. Certain constructs were added to standard PEARL, e.g., enhanced tasking operations and real-time specific programming support as described in the previous section.

Strict temporal determinism on the programming language level can only be achieved under the assumption that each statement is executed in a bounded and a priori predictable time period. For this reason, each feature that could take an arbitrarily long time to execute is either renounced in miniPEARL or is guarded with a timeout clause. For similar reasons, there is no recursion and dynamic data structures are not used. Also, maximum numbers of loop repetitions are known at compile time.

Since the temporal execution behavior of a program as a whole is tightly correlated to that of all its parts, all program code must be compiled at the same time. Although the organization of programs is modular, modules may not be compiled separately (except for syntax checking). First, the compiler transforms an application's source code into an intermediate form, in the course of which a modified syntax tree is constructed and symbol tables for identifiers, procedures and tasks are built. To this internal form of a program global code optimization is applied. Then, code generation and timing analysis are performed. They both use so-called translation macros. These macros are specific for the hardware architecture and the processor used. To support different platforms, different sets of macros are required.

Each element of a syntax tree corresponds to one or more macros. As an example, a macro that corresponds to the addition of two integer variables is shown on Figure 5.

In this macro, the placeholders &1 and &2 are replaced by the first and the second calling parameter, respectively. Output code is generated in the code generation phase either in assembly or in machine language form. In the first case, the generated code must be assembled before it is loaded into the target system. In the second case, compiling is quicker, but the compiler output is not readable and, thus, difficult to verify.

The same macro is also used in the timing analysis phase. The directive TIME states the basic execution time of this macro's code. It is not implemented as an attribute of the macro because, for code optimization, conditional expansion of macros is allowed and execution times may vary with operand types or values. The exact execution time will be established when the macro is called by the compiler and operands are given. For the example macro, its call could look like:

ADDI L(4),3

where the notation L(4) represents an absolutely addressed local variable residing in location 4 and the second operand is a constant. Additional time used for effective address calculation and operand fetches and stores is added yielding the total execution time of the macro.

Since the language is relatively simple, it can be more efficiently compiled into object code, and the latter more easily optimized. Moreover, the analysis of the code's temporal execution behavior is easier, more accurate and less pessimistic. Through accessing compiler internal data structures, the syntax tree can be precisely analyzed, and any global or local program code optimizations considered.

Pessimism in task run-time estimations. A drawback of many methods for task execution time estimation, and one of the reasons why they are not more widely used, is that they yield relatively pessimistic results. The reasons can be found on all layers involved in programming, from non-deterministic language features via operating system overhead to hardware behavior.

In our approach, the influence of the operating system overhead was minimized by migrating the kernel to a separate processor. Only context-switches and non-pre-emptable critical regions have to be considered.

State-of-the-art processors are optimized towards average performance and include features that behave, as a rule, temporally non-deterministically, like caching, internal parallelism, pipelining etc. These features cannot be consistently considered in temporal analysis. Often, worst-case execution times are given, and sometimes even statistically estimated average times, only. Instruction execution times are heavily dependent on their sequential order in program code. Animating sequences of instructions often fails, because their exact execution behavior is a proprietary information.

In our concept of macros it is possible to see whether a preceding macro does not end with a jump or branch enabling the subsequent one to start earlier with a pre-fetch. A macro can, thus, give the subsequent one a "bonus", which may be subtracted from its execution time.

For even more precise run-time determination, our analyzer provides another option — generation of pilot code. The idea behind is to measure execution times of actual code on a target system. Since a target system may not be available at the time of software development, and since it is not always easy to set the worst-case test scenario, the compiler is used to generate pilot object code. This is characterized by always running through the path with the longest execution time (the longest alternative in IF and CASE statements is always selected, the maximum possible number of loop iterations is forced, etc.). Further, input/output commands are replaced by corresponding delays. Thus, only partially built or similar target systems with the same processor are sufficient for the measurements.

Most problems, however, are introduced by programming itself. Studies (cp., e.g., (Puschner and Koza, 1989; Park, 1993)) based on supplying additional knowledge about programs to analyzers were made to achieve better estimation of run-times. Although yielding very good results that are close to the actual values, these solutions do not appear to be sufficiently practical for implementation, yet. If application designers feel that an automatic estimation is excessively pessimistic, they can thoroughly test and measure the critical parts of a given code until they become familiar with the temporal circumstances. In this case they can make use of the possibility to explicitly define execution times for such code segments and, thus, override automatic estimation.

5.   Typical Set-up and Design of Applications

As an example, let us consider our embedded computer controller to be built into a classical industrial process consisting of mechanical production machines. Typically, such a process is physically distributed over a production hall. The control system is sensing system states, characteristic values, and data inputs and, with calculated results, controls the actuators. The times to react upon events from the process are in the order of magnitude of milliseconds and must be guaranteed.

Sensors provide two kinds of data: they acquire values of input variables and notify the control system of events, which influence the further process behavior (e.g., changes of certain internal states like start or termination of certain activities, operator interventions, alerts, etc.). The former are polled by the intelligent peripheral interfaces, and controlled by task processors synchronously with application program processing. The events have the form of asynchronous signals, and are directly fed to the kernel processor's primary reaction layer. Upon occurrence of these signals, of programmed time events and changes of synchronizers and common variables, unified system events are reported to the secondary reaction layer, where associated tasks are then scheduled earliest deadline first. If necessary, a context-switch is requested in the corresponding task processor to activate a task, handling an event. Such tasks may request certain operating system kernel services during their execution.

The intelligent peripheral interfaces are selected to be mounted on the objects from which they acquire input data and/or in which they drive actuators. The associated application tasks are statically mapped onto one to four task processors, depending on the performance required.

The application tasks are programmed in miniPEARL, which represents a simple, deterministic, and, to certain extent, safe (Colnarič, Verber and Halang, 1995) subset of a standardized high-level programming language, convenient for control applications. The particular hardware configuration is described in the system part of the application program. Then, the application or problem part is designed, containing the actual algorithmic implementation. The syntax of miniPEARL already includes all necessary scheduling constructs and other operating system calls; the application programmer does not have to be aware of specific operating system issues.

Computer aided application engineering is not supported, yet. It is one of the directions for further research which will be commenced next year, and which will feature aspects like graphical programming, specification animation, automatic program generation, rapid prototyping, and stepwise refinement of functional and temporal issues. An interesting and motivative approach towards the design of hardware configurations and software systems, however based on DSPs, can be found in (Lauwereins et al., 1995).

6. Conclusion

In order to assure predictable behavior of real-time systems, it is necessary to determine a priori bounds for task execution times. In this paper a consistent holistic design of an experimental control system for embedded applications operating in the hard real-time domain is described. It is shown how the execution times of tasks running on task processors can be estimated, providing the necessary information for schedulability analysis.

In this project, we did not intend to develop many novel concepts. We merely tried to strictly apply known ones and to combine different design domains to obtain practically usable experience in designing predictably behaving embedded control systems.

This report is an intermediate one: we are concluding the design of the compiler, whereas the hardware platform and the operating system are already in the system integration phase. Due to some radical changes in the design of the hardware platform, the compiler and the run-time analyzer had to be re-designed. For this reason we presently cannot comment on the quality of run-time estimation, nor assess the practicability of program execution time measurement on the pilot system. The performance of the system is not expected to be surprisingly high, but it can and will be optimized in the next versions. After integration and final testing we are planning experimental applications in robotics and industrial process control to carry out practical evaluations.

The proposed approach involves many components; its price exceeds the price of standard industrial controllers. Although the hardware price is a significant parameter, it is still low in comparison to the costs of an application environment and of software design. Furthermore, a consistently designed control system can prevent costs of production standstill and/or damages, or even endangerment of human health or lives, caused by failures, which can by far exceed the hardware costs.

We believe that by designing embedded control systems in this way the requirements of hard real-time systems, viz., timeliness, simultaneity, predictability, and dependability, will be better met than by the prevailing state-of-the-art.

References

Blume, W. and Klinker, W. 1993. *The Sensor/Actuator Bus: Theory an Practice of Interbus-S*. Landsberg/Lech: Verlag Moderne Industrie - Die Bibliothek der Technik, Band 78.

Colnarič, M., Halang, W. and Tol, R. 1994. Hardware supported hard real-time operating system kernel. *Microprocessors and Microsystems* 18:579–591.

Colnarič, M., Verber, D. and Halang, W. 1995. Supporting high integrity and behavioral predictability of hard real-time systems. *Informatica, Special Issue on Parallel and Distributed Real-Time Systems* 19:59–69.

Cooling, J. 1993. Task scheduler for hard real-time embedded systems. *Proc. of International Workshop on Systems Engineering for Real-Time Applications*, pp. 196–201, London: IEE.

Halang, W. 1988. Definition of an auxiliary processor dedicated to real-time operating system kernels. Technical Report UILU-ENG-88-2228 CSG-87, University of Illinois at Urbana-Champaign.

Halang, W. and Stoyenko, A. 1991. *Constructing Predictable Real Time Systems*. Boston-Dordrecht-London: Kluwer Academic Publishers.

Halang, W. 1993. Achieving jitter-free and predictable real-time control by accurately timed computer peripherals. *Control Engineering Practice* 1:979-987.

Henn, R. 1989. Feasible Processor Allocation in a Hard-Real-Time Environment. *Real-Time Systems* 1:77–93.

Lauwereins, R., Engels, M., Adé, M. and Peperstraete, J. 1995. Grape-II: A system-level prototyping environment for DSP applications. *IEEE Computer* 28:35–44.

Lindh, L. 1989. Utilization of Hardware Parallelism in Realizing Real-Time Kernels. PhD thesis, Royal Institute of Technology, Sweden.

Park, C. 1993. Predicting program execution times by analyzing static and dynamic program paths. *Real-Time Systems* 5:31–62.

Puschner, P. and Koza, C. 1989. Calculating the maximum execution time of real-time programs. *Real-Time Systems* 1:159–176.

Roos, J. 1990. The performance of a prototype coprocessor for Ada tasking. *Proc. of the Conference Tri-Ada*, pp. 265–279.

Stankovic, J. 1988. Misconceptions about real-time computing. *IEEE Computer* 21:10–19.

Stankovic, J. 1995. Adjustable flow control filters, reflective memories, and coprocessors as support for distributed real-time systems. *International Journal of Mini and Microcomputers* 17.

Stankovic, J. and Ramamritham, K. 1990. Editorial: What is predictability for real-time systems. *Real-Time Systems* 2:246–254.

Stoyenko, A. 1987. A Real-Time Language With A Schedulability Analyzer. PhD thesis, University of Toronto.

Verber, D. and Colnarič, M. 1993. *Proc. of Software Engineering for Real-Time Applications Workshop*, pp. 166–171, London: IEE 1993.

Verber, D. and Colnarič, M. 1995. Programming and Time Analysis of Hard Real-Time Applications. *WRTP'95, Preprints of IFAC/IFIP Workshop on Real-Time Programming*, Fort Lauderdale, Florida.

Real-Time Systems, 14, 311–323 (1998)

HEDRA: Heterogeneous Distributed Real-Time Architecture

H. THIELEMANS   hans.thielemans@mech.kuleuven.ac.be

L DEMEESTERE

H. VAN BRUSSEL   hendrik.vanbrussel@mech.kuleuven.ac.be

Katholieke Universiteit Leuven, Dept. of Mechanical Engineering, Div. PMA, Celestijnenlaan 300B, B-3001 Heverlee, Belgium

Abstract. HEDRA, ESPRIT project (nr. 6768), aims to develop a heterogeneous distributed real-time architecture for robot and machine control. This paper describes an open and flexible programming system, as a part of this architecture, that is based on an existing system called VIRTUOSO. The programming system is an evolution of Virtuoso towards the aimed architecture. The work concentrates on the achievement of guaranteed real-time behavior, minimum interrupt latency and transparency of interprocessor data communication.

Keywords: distributed control, real-time systems, parallel processors, multi-processor systems, programming environments

1. Introduction

Manufacturing systems often consist of several functional modules that have to cooperate. With decreasing processor cost it is often advantageous to have specific computing power placed where it is needed. A broad range of processors can be used each having a typical field of application. Sometimes it is desirable to benefit from processor specific characteristics to optimize the overall system performance. This results in distributed systems that contain a heterogeneous set of dedicated processors. The use of multi-processor systems, specially those containing a mix of processor types, is normally avoided due to their complexity. Existing multi-processor systems are mostly found in a master-slave configuration: a master processor and several slave processors with limited interaction.

Several real-time systems exist that can be used in manufacturing systems. Some of them are commercially available while others are the result from research projects (Lee et al. 1988; Clark, 1989; Stewart et al. 1990). In general existing real-time systems suffer from one or more shortcomings: no multi-processor capabilities, insufficient interprocessor communication and no support for heterogeneous systems.

HEDRA aims to remove the limitations for the use of heterogeneous multi-processor systems. As a result an open and flexible development and operational environment has been created which can be used for the development of real-time control applications.

HEDRA originates from an existing real-time programming system called VIRTUOSO. Real-time kernel and development environment have been extended and have been ported to HEDRA. Typical problems of the proposed architectures like the data representation on different processor types and system configuration are tackled in the project.

The HEDRA architecture has been applied for the control of a bending workcell. The controller architecture consists of a mix of transputers and DSPs TI320C40.

2. Virtuoso: The Basis of HEDRA

The VIRTUOSO real-time programming system has been chosen as a basis for the HEDRA architecture. The programming system contains a multi-processor real-time kernel as the operational part and an application development environment. Virtuoso is a commercial programming system from the Belgian company Eonic Systems, nv. The real-time kernel from Virtuoso is based on the RTXC kernel for PC that was ported to the transputer (Verhulst et al. 1990a; Verhulst et al. 1990b; Thielemans et al. 1991). Originally the kernel was a single-processor kernel, later on, due to the transputers parallel processing capabilities, it has been expanded to a multi-processor real-time kernel. Virtuoso has been ported now to several environments. It supports a broad range of processor families (Motorola 68xxx, transputer, DSP TI320C30 and C40, AD21020,...).

The main reasons for the selection of Virtuoso were the support for the available and selected hardware, the ease of expansion and the previous experience with the product. Other vendors' multiprocessor real-time systems are mostly limited to shared memory and ethernet communication.

The philosophy of Virtuoso is to put a software shield on top of complex hardware systems protecting the application level from processor specific hardware. By shielding the application from the hardware level it creates a virtual uni-processor environment. For the developer it is as if he is working with one big (virtual) processor.

2.1.   The Virtuoso/HEDRA real-time kernel

The real-time kernel is a multi-tasking pre-emptive kernel that uses a priority-driven scheduling algorithm. Timely execution of applications reflects the relative priorities of the processes they are mapped on. The kernel provides basic functionalities required for guaranteed real-time behavior by means of kernel objects:

- fast kernel operation and context switching,
- multi-tasking with multiple-priority processes,
- process model: executing thread, static process stack and process communication port,
- real-time clock with adjustable tick frequency,
- static memory allocation,
- process synchronization with timers and semaphores,
- interprocess communication with queues, mailboxes and ports,
- priority inheritance (Morgan, 1993),
- device support with device drivers and resources.

Priority-based waiting lists within the kernel for kernel object operation (e.g., waiting on a message in a mailbox) and priority-driven communication over the processor network guarantee short response times for high priority processes.

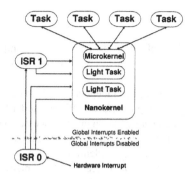

Figure 1. Multi-level structure of the real-time kernel.

A message based protocol is used for communication with the kernel (for service requests) and interprocessor operation (for remote kernel calls). If a kernel call is identified for execution on a remote processor, the local kernel will invoke a remote call. It will pack the user message into a command message and send it to the remote processor. The system router will route the message to the appropriate processor through the processor network. The remote processor will unpack the message and execute it.

The use of a kernel imposes computational overheads, introducing important restrictions on applications. The open multi-level structure of the Virtuoso real-time kernel can avoid these problems. The user can implement his code at four different programming levels. The multi-level structure is represented in figure 1. Each level has its own implementation characteristics and limitations:

Level 1 - ISR0 level: Interrupts are accepted. They can be handled completely at this level or passed to a higher level. During operation all interrupts are disabled. Execution times for accepting an interrupt are less than one μsec.

Level 2 - ISR1 level: Interrupts are normally handled at ISR1 level. Execution at this level proceeds with interrupts enabled. Other interrupts can be accepted during the handling of an interrupt.

Level 3 - nanokernel level: Nanokernel processes, running at this level, have a reduced context, allowing submicrosecond context switching. They are used for time-critical applications and the implementation of device drivers.

Level 4 - microkernel level: This is the application level with priority-driven and preemptive operation. At this level, computational drawbacks of the kernel operation are the most observable.

In these circumstances the system can keep up with high sampling rates even during extensive kernel operation.

An important advantage of the Virtuoso real-time kernel is the portability of the kernel code. The major part of the kernel is written in a high-level programming language (C). Only a minor part, the time-critical code, is processor specific and often optimized in assembler.

2.2. The Virtuoso/HEDRA API

Kernel objects accessible to the application level can only be manipulated through the kernel level. Sending user messages to the kernel, applications can invoke a request for object manipulation.

To allow application code to be written independent of the hardware the Virtuoso kernel provides an application programming interface (API), a standard set of kernel calls, which is hardware independent. Similar semantics on every processor type makes the application code portable to any other hardware platform without modification.

Kernel objects have a system-wide unique identifier indicating the processor the object is residing on and the object number on that processor. The object identifier is passed to the kernel during a kernel call.

2.3. System services

A standard set of system services is available through the use of a host system:

- File I/O,

- Graphical I/O,

- Keyboard I/O.

On the host system a server task is running, communicating with the processor network. Driver processes on the root processor forward service requests to the server process on the host system which executes them.

2.4. Application development

The Virtuoso programming system provides a virtual uni-processor environment for the design of applications. The user can write the application code as if he is working on one big processor. This offers some advantages:

- application code is independent of the processor the code will finally run on.

- hardware decisions do not influence the software system. Decisions like hardware architectural layout can be postponed, so that the hardware architecture can meet software requirements.

Design and implementation lead to the definition of a set of kernel objects (tasks and cooperation kernel objects like queues, semaphores). Kernel objects are used in the application code using their symbolic name.

After the application development, the user assigns the application to a hardware system during the configuration phase, the binding of symbolic kernel objects to the physical system. Kernel objects are allocated to processors and their physical properties are defined. Figure 2 gives an example of an application and the mapping on a physical system.

All configuration information is retained in a configuration file, a textual representation of software and hardware configuration and some kernel configuration parameters:

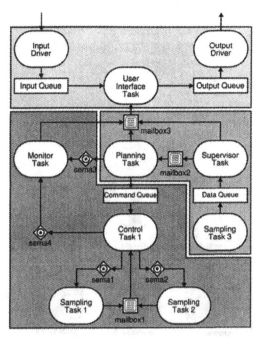

☐ Processor 1 ▦ Processor 2 ■ Processor 3

Figure 2. Example of an application.

- software configuration: definition of kernel objects and their physical properties;

- hardware configuration: definition of processors, interprocessor links, devices;

- kernel configuration parameters: mail packets, data packets, real-time clock.

A system generation tool (sysgen) then generates all processor dependent files. It calculates routing tables for shortest path routes and generates source code files. Figure 3 gives an overview of the system configuration.

   The Virtuoso programming system provides some tools supporting the user during system profiling:

- distributed debugger: the distributed debugger checks the status of the kernel (objects) during operation for every processor in the network.

- distributed monitor: gives an overview of the kernel events and the processor time needed for the execution of processes and kernel calls.

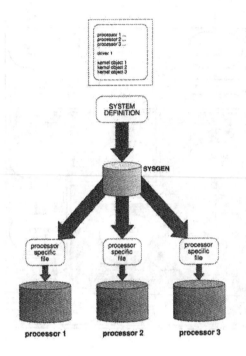

Figure 3. System configuration.

3. Porting Virtuoso to HEDRA

The Virtuoso real-time system serves very well the objectives of HEDRA. Porting Virtuoso to an architecture HEDRA is aiming at, reveals some typical problems.

The major interest of HEDRA is a heterogeneous architecture. In such a heterogeneous system the consistency of shared data between processes running on a different processor becomes an important topic during the allocation of processes to the available processors.

3.1. Shared data in heterogeneous systems

Various processor types usually have a typical processor architecture. Special processor functionalities and commercial benefits generally result in different designs incompatible with other processor families. For communicating processors, this imposes a considerable obstruction for the introduction of heterogeneous multi-processor architectures in control systems. Important communication problems in a heterogeneous system are:

Incompatible communication interfaces. Every processor has different facilities for interprocessor communication, mainly determined by the processor cost and the processor package. Processor communication systems range from serial point-to-point interconnections, to high-speed parallel bus systems. Incompatibility of the communication system is generally two-fold: an electrical incompatibility, and the nature of the communication itself. In the application involved, the communication between the transputers and the DSP's is done using serial transputer links at 20 Mbit/s, using special hardware at the DSP side to convert to/from a DSP link. The DSP's have among them serial channels at 20Mbyte/s. The smallest communication entity for the DSP is one word (4 bytes), while the transputer allows one byte. This incompatibility is solved in software.

Incompatible internal data representations. The internal data representation is the way a processor handles and stores its basic data types or data structures. Two elements determine the data representation: the target processor and its compiler. Every processor offers a limited set of hardware solutions for data representation, called "primitive data types". Although most processors support a similar set of primitive data types (e.g., int or double in the C language), their representations can be quite different, causing the main problem for interprocessor communication in a heterogeneous system. The difference between little and big endian formats, or Intel versus Motorola processors is also a typical example. The compiler maps data types of the programming language on the set of primitive data types. The description of the internal format of data types has not been standardized, making compiler differences possible.

Loss of precision. Due to different lengths of data type representations, communication can involve loss of precision. Doubles on a transputer T8 have 64-bit precision while, they only have 32-bit precision on a DSP TI320C40. On the other hand, a different representation length can cause data corruption of adjacent data in the processor memory, when data structures of a different length are shared.

HEDRA conversion support. Within the philosophy of HEDRA, a virtual uni-processor system, migration of processes should be possible with minor system modification, even over boundaries of processor families. A solution often used in heterogeneous systems is to translate data to a canonical format before communication (Jones et al. 1992), as in SUN's Open Network Computing (ONC) system. The real-time kernel on every processor has an instance of a conversion library for the translation of the native format to the canonical format. In a network with a limited set of processor families, this frequently leads to useless conversions since communication between processors of the same type is likely. This induces overheads for every communication operation.

In the HEDRA communication system the Network Computing Architecture (NCA) of Apollo/HP is followed. This uses the *receiver-makes-it-right* approach. Interprocessor communication is in the native format of the sending processor. The receiving kernel decides on conversion, based on the processor type of the sender and the contents of the message. This communication system offers some advantages:

- There is only a minimal conversion overhead for interprocessor communication.

- Accuracy problems of transmitted data can be minimized.

- The communication system is flexible. Either the system or the application can decide on the need for conversion, and on the type of conversion (full, partial or special conversion).

For every set of processor types, two conversion libraries are required but with increasing memory capabilities and the limited number of processor types this need not cause serious problems.

To allow a generalized calling mechanism to conversion functions, applicable at both the user and the system levels, a structuring and implementation scheme is used for the development of the conversion libraries.

When applying fully automatic conversions, a communication preprocessor scans the source code, registers all communication operations and data structures involved in communication, and finally modifies the source code where needed. The information about the contents of the data structure is written in a type structure and stored in the source code. These type structures are used for the on-line data conversion. This way the user achieves fully transparent data conversion.

3.2. *Development tools for heterogeneous systems*

Development tools of the programming system, like the distributed debugger and system configuration tool are modified to support application development for heterogeneous systems.

A new version of the configuration tool, with graphical user interface and support for resource allocation, is under development.

The host server is converted to a multi-threaded process running under the OS/2 or UNIX operating system. The server is integrated in the processor network not distinguishable from other processors. The host will run a reduced version of the kernel supporting basic functionality like network communication. It increases the portability of system services from host to processor network and inverse.

3.3. *Configuration of an open and flexible controller*

In industrial environments, hardware reconfigurations are very likely. During test phase or operation, wiring of input and output connections from the controller to the environment can change frequently. These modifications should be possible without a complete recompilation of the control software. Within HEDRA the I/O configuration is contained in an I/O description file. During system initialization, the information about the I/O lay-out is read from this file. Modification of device drivers is under development to support the I/O initialization.

4. Application

HEDRA is a project within the ESPRIT program of the E.U.. The project consortium consists of 3 industrial partners, BARCO Industries N.V. (B), LVD Company N.V. (B) and HEMA Elektronik GmbH (D), and 2 universities, Universität Stuttgart Institut für Steurungstechnik (ISW) (D) and K.U.Leuven Division Production Engineering, Machine Design and Automation (PMA) (B). Industrial partners mainly concentrate on the development of hardware related components, universities on the other hand work on software developments.

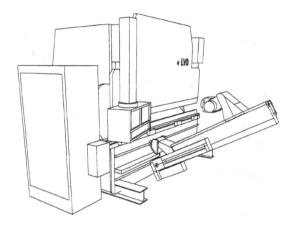

Figure 4. HEDRA demonstrator. A press brake with a cooperating metal-sheet-handling robot.

4.1. Application hardware

The HEDRA architecture has been applied for the controller of an industrial manufacturing cell, the final demonstrator of the project (see Fig. 4). The workcell, which is an LVD machine, consists of a multi-axis numerically controlled bending machine and an integrated 5-axis bending robot. The cooperating bending robot follows and supports the metal sheet during bending. The robot control algorithm compensates model inaccuracies and tracking errors based on the measured bending angle and forces in the robot wrist. A force sensor, mounted on the robot, measures forces at the robot wrist. A measurement system, mounted on the press brake, gives an on-line evaluation of the bending angle of the metal sheet. In addition the robot manipulates sheets before and after bending: loading and unloading of the machine and repositions the metal sheet between two bending sequences.

The prototype workcell controller architecture consists of a mix of transputer (T2 and T8) and DSP (TI320C40) processors. Both processor types have excellent provisions for interprocessor communication. Distinction is made between several types of processor nodes: pure processor nodes for calculations, graphical nodes for display purposes and I/O nodes for interface to the work-cell (see Fig. 5). For the I/O nodes the VME bus is used as a fast I/O communication system, mainly for sensor and actuator level communication. Each I/O related processor node has an interface to the VME bus. Standardization on the industrial VME bus has several advantages:

- integration of multi-vendor processor and I/O boards in one architecture,

- distribution of the I/O functionality over multiple processors nodes (with the same I/O capabilities),

- I/O boards with the related processes, like device drivers, can be moved to another processor node.

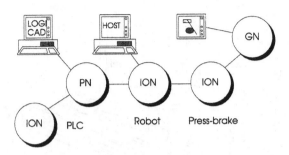

Figure 5. Demonstrator Hardware Nodes

BARCO and HEMA, both active in the field of control equipment, were responsible for the development of the required electronic hardware.

HEMA has developed transputer and DSP processor boards, both with hardware interface for interprocessor communication and hardware interface to the VME bus.

BARCO and LVD have developed an intelligent multi-axis I/O board capable of controlling up to 12 axes. The board is equipped with a Motorola micro-controller that provides data conditioning and all low-level functionality, like data sampling and output updating. This reduces overhead for computing processors and data transfers over the I/O bus.

4.2. Application software

The controller software, running into the HEDRA architecture can be divided in several functional modules. Figure 6 represents the structure of the application software.

- MMI : The MMI module provides the functionality for the interaction with the operator. Basic functions are application programming, operator interface, manual machine control, display and diagnosis functions

- NC/RC Core: The NC/RC Core contains the functionality for movement generation of press brake and robot. Modules that can be distinguished in NC/RC Core are: interpreter, interpolator and transformation.

- Axis Control: The lowest-level control, this runs at a fixed sampling rate with the most stringent real-time constraints.

- Process Control: The process control module contains functionalities to control the manufacturing process (e.g., corrected position for the robot gripper).

- PLC: The Programmable Logic Controller is a logic control module that combines input signals and sets output signals. It is implemented as a software module in the controller and replaces the conventional hardware PLC module.

The sequence of operations during a bending operation is controlled through messages between the different controller modules (Fig. 7). The robot takes a plate from a supplier

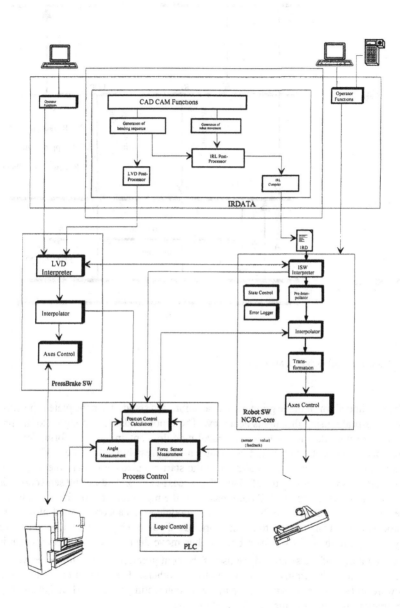

Figure 6. Structure of the application software

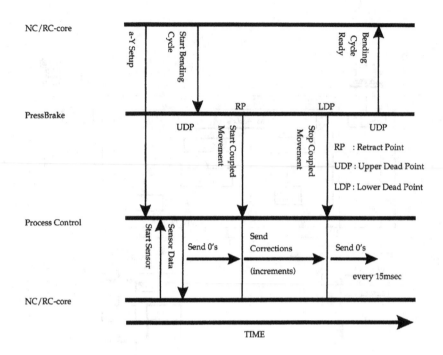

Figure 7. Module Synchronization Messages

mechanism and aligns it in the gripper using a calibration stand. Then the plate is positioned in the press-brake using the robots accuracy. The robot sends a setup signal to the press-brake control module and the process control module, containing the information on the desired bend. Then it gives a start command to the press-brake module. When making contact with the plate the press-brake sends the start-coupled-movement message. From this moment, at every sample-time (15 ms) the process-control module will send corrections for the robot joints to the NC/RC core, based on the measurements of force and bending angle. At the lower dead point the press-brake module sends a stop-coupled-movement to the process control and retracts to the the upper dead point. Then it sends a bending-cycle-ready message to the NC/RC. The robot can then move the plate to a delivery mechanism.

The modularity of the setup and the use of the configuration tools, allowed easy testing of each module with dummy replacements for the others. The integration of all modules (hardware and software) was then just a matter of connecting the network and changing the configuration file, recompiling, linking and loading.

ISW was responsible for the implementation of the robot control software. Existing press brake software from LVD, running on a Motorola based controller, was ported to the new architecture. PMA was the responsible partner for the design and redesign of Virtuoso towards HEDRA.

5. Conclusion

Distributed multi-processor systems with mixed processor types emerge in controllers for manufacturing systems. The lack of development tools however slows down their introduction. HEDRA presents a real-time programming system for heterogeneous multi-processor architectures that will ease the development of industrial control systems consisting of mixed processor types.

Acknowledgments

The work was sponsored by the HEDRA project of the EUROPEAN Union, within the ESPRIT programs (ESPRIT 6768) and the Belgian Programme on Interuniversity Attraction Poles initiated by the Belgian State – Prime Ministers's Office – Science Policy Programming (IUAP-50).

References

Clark, D. 1989. HIC: An Operating System for Hierarchies of Servo Loops. *Proc. Int Conf. on Rob. and Automation, 2*, pp. 1004-1009.

Jones, J., Butler, P., Johnston, S. and Heywood, T. 1992. Hetero Helix: Synchronous and asynchronous control systems in heterogeneous distributed networks. *Robotics and Autonomous Systems* 10:85-99.

Lee, I., King, R. 1988. Timix : A Distributed Real-Time Kernel for Multi-Sensor Robots. *Proc. of IEEE Int. Conf on Rob. and Automation, 3* pp. 1587-1589.

Morgan, K.D. 1993. The HP-RT Real-Time Operating System. *Hewlett-Packard Journal* 8:23-30.

Stewart, D., Schmitz, D. and Khosla, P. 1990. Implementing Real-Time Robotic Systems Using CHIMERA II. *Proc. Int. Conf. on Rob. and Automation*, pp. 598-603.

Thielemans, H., Verhulst, E. 1991. Implementation issues of TRANS-RTXc on the Transputer. *IFAC Algorithms and Architectures for Real-Time Control*, pp. 123-128.

Verhulst, E., Thielemans, H. 1990(a). Pre-emptive scheduling and meeting hard real-time constraints with TRANS-RTXc on the Transputer. In *Application of Transputers, 2*, pp. 288-295, IOS Press.

Verhulst, E., Thielemans, H. 1990(b). Predictable response times and portable hard real-time systems with TRANS-RTXc on the Transputer. In *Real-Time Systems with Transputers*, pp. 232-240, IOS Press.

Real-Time Systems, 14, 325–343 (1998)

Open Embedded Control

KLAS NILSSON   klas.nilsson@dna.lth.se

ANDERS BLOMDELL   anders.blomdell@control.lth.se

OLOF LAURIN   olof.laurin@adra.se

Dept. of Automatic Control, Lund Inst. of Technology, P.O.Box 118, S-221 00 Lund, Sweden

Abstract. Embedded control devices today usually allow parameter changes, and possibly activation of different pre-implemented algorithms. Full reprogramming using the complete source code is not allowed for safety, efficiency, and proprietary reasons. For these reasons, embedded regulators are quite rigid and closed concerning the control structure.

In several applications, like industrial robots, there is a need to tailor the low level control to meet specific application demands. In order to meet the efficiency and safety demands, a way of building more generic and open regulators has been developed. The key idea is to use pieces of compiled executable code as functional operators, which in the simplest case may appear as ordinary control parameters. In an object oriented framework, this means that new methods can be added to controller objects after implementation of the basic control, even at run-time.

The implementation was carried out in industrially well accepted languages such as C and C++. The dynamic binding at run-time differs from ordinary dynamic linking in that only a subset of the symbols can be used. This subset is defined by the fixed part of the system. The safety demands can therefore still be fulfilled. Encouraged by results from fully implemented test cases, we believe that extensive use of this concept will admit more open, still efficient, embedded systems.

Keywords: embedded systems, open systems, real-time systems, dynamic binding

1. Introduction

Making machines programmable has added flexibility to manufacturing equipment. Typical examples are industrial robots and NC-machines. They can be reprogrammed for applications that were not defined at the time the equipment was developed. Improved functionality/performance and cost efficiency are other major driving forces for use of computer control, for instance in vehicles. Reprogrammability is also very desirable in this case, which means that the lifetime of the product can be prolonged without having to exchange the components when new (e.g., environmental) demands are imposed.

In mechatronics and process control, demands on safety, efficiency and simplicity are often contradictory to flexibility. Products providing features for end-user programming, such as industrial robots, are therefore preferably layered in such a way that the ordinary user can do the application programming in a simple and safe way, and the expert programmer can tailor the system to fulfill unforeseen requirements from new applications. The same also applies to many mechatronic products that do not provide end-user programming. For example, the multivariable engine control in a car should be possible to update by the car manufacturer. Furthermore, within mechatronics it is in general desirable to use components that can satisfy the needs of many applications and products. This will contribute to cost reduction, which means that we have the same situation as for the robot control mentioned above; *open components* are needed.

A system that allows such low-level reconfiguration to meet requirements that it was not specifically designed for, is called an *open system*. Most machine controllers used in industry today are *closed*. The reason is that they have been specifically designed to provide an easy to use, cost efficient, and safely performing control for today's standard applications. In the present and future more advanced applications, however, customization of closed systems becomes time consuming and thereby costly. Development towards more open control systems is therefore ongoing, refer to (Nilsson and Nielsen, 1993; Ford, 1994) and references therein. Still, there are proprietary, efficiency, and safety reasons for not having a completely open system. The key issue is to find the right tradeoffs between an open and a closed design of a system. The best possible solution will, of course, depend on the available software mechanisms. We include both (partly) open systems and components in the term *open embedded control*. The term control is used since the purpose of behaviors subject to change are typically to control the system, or parts of it.

The trend towards more open systems has been ongoing for some time within operating systems, computer networks, interactive software, and graphical user interfaces. This development has resulted in software techniques that have inspired our approach to open embedded systems. We do, however, not want to impose the use of standard (heavy) operating systems and process communication principles on an otherwise lean design of a cost-sensitive mechatronic product.

The desire to create software mechanisms that support simple implementation of open embedded control can now be formulated as: Given appropriate hardware, how should the software be organized to allow efficient extension (in terms of improved functionality, performance, etc.) by changing open parts of the software only, while some parts of the system may remain closed. As the examples in the next section will show, only allowing parameter changes and use of pre-implemented algorithms does not provide enough flexibility.

One approach, used in PLC systems, is to down-load new code for interpretation to each hardware unit. In many applications, full utilization of the computer hardware is required, and executable code (cross-compiled to the specific hardware) is therefore preferred, and also used in a modern process control system like SattLine (Johannesson, 1994). Use of compiled libraries admits the implementation of certain components to be hidden, but often we need closed subsystems with open 'slots'. Recall, however, that we want to keep part of the software closed. The question is then how to obtain a suitable border between the open and closed parts. This paper presents a solution to this problem, without requiring any special language or run-time model. We therefore believe that the proposed way of building open real-time control systems is an innovative simplification compared to more general concepts (like (IBM, 1993; Object Management Group, 1991; Microsoft, 1993; Apple, 1993; Sun, 1993; Dorward and Sethi, 1990)) for open systems.

2. Applications

Some typical situations when an embedded system needs to be open are presented in this section. These examples are all in the area of manufacturing systems, but they also capture typical properties of many other types of embedded systems.

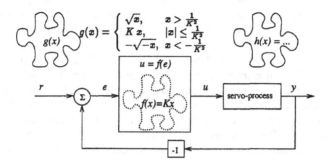

Figure 1. Block diagram of a P-controller allowing the gain to be replaced (at run-time) by a nonlinear function. The puzzle pieces illustrate type-checking during dynamic binding.

2.1. Feedback control

The most common type of feedback controller in industry is the PID-regulator (Åström and Hägglund, 1995). In most process control applications, PID-controllers give sufficient performance and there are simple rules and schemes for how to do the tuning without requiring extensive knowledge about the process dynamics. Although the basic PID algorithm is very simple, industrial application demands have resulted in enhancements like gain-scheduling, auto-tuning, and adaptivity. Such features have been built into more advanced PID regulators.

Within mechatronics and machine control, ability to compensate for known nonlinearities is often more important for performance than the ability to handle unknown changes in process dynamics (called adaptivity above). Compensation of nonlinearities in the process typically results in nonlinear control functions. Such functions are normally specific for the application. Nevertheless, like in the process control case, it is desirable to use standard control blocks in terms of hardware and software. This is usually not possible within mechatronics today, but such flexibility concerning the embedded control functions could reduce the cost of the controller in many cases. Furthermore, the sampling frequencies required in mechatronics are typically much higher than those used in process control. Special care has to be taken in order to achieve the flexibility in an efficient way.

As a simple example of industrial motion control, assume we want to control the motor angle in a mechanical servo system. Further assume that we do not accept any overshoot when the desired angle r is approached, and that an inner analog velocity controller is used together with drive electronics that controls the motor torque within a certain range. This means that we have a servo process (see Figure 1) with an inner nonlinearity due to the limited torque available. Only the P-part of the PID controller is used in this case.

With some basic knowledge in physics and control, one can easily see that the nonlinearity can be almost fully compensated for using the function $g(x)$ shown in Figure 1. The trick is to use the square-root function to compensate that the required braking energy is proportional to the square of the speed, a fact that should be well known to all car drivers. For stability reasons, the function g contains a linear part for small arguments. This type of nonlinearity is used in some industrial drive systems, but implemented as a built-in special purpose feature.

If we instead make an embedded PID-controller in which the proportional gain constant can be exchanged by an external function, much greater flexibility is achieved. These functions need not be known when the controller is shipped. They can be written and inserted when the controller is to be used. In the next section, we will return to how this can be achieved, but it means that we can use the same embedded controller as a building block for controlling very different processes. Type checking ensures that only a conformant function can be inserted. The type of the functions is illustrated in Figure 1 by the shape of the pieces.

It is common within mechatronic applications to make use of a dynamic model of the controlled system to achieve the best possible performance. Control techniques like state feedback, state observers, optimal control, linearizing control, etc., are used for this purpose. Control of manufacturing equipment and many military control problems are typical applications for such control strategies. In such systems, like industrial robots, it is even more valuable to have software slots admitting the behavior of the machine to be tailored to different needs.

2.2. Supervisory control

A distributed control system is composed of separate local controllers and at least one supervisory controller. In such a system it is preferable to have a vendor independent interface to the local units. It is also important to minimize data flow and reduce the computational load on the supervisory controller. However, standard controllers for local use have different signals available. At construction time of a controller it is impossible to know which signals might be desirable to use, since this depends on the specific application. In other words, distributed control systems today are sometimes burdened with more communication and computations than necessary.

An attractive solution would be to have a slot in the local controllers for additional computation. This could be in the form of algorithms on a smartcard supplied by the vendor or executable code loaded over a local network. With this approach standard controllers could be customized to fit different environments, and the computational load on the supervisory controller could be reduced.

This type of functionality partially exists in today's process control systems, but requires that the local controller is integrated with the supervisory control which either sends down new code for interpretation, or restarts the local controller with new compiled code. Dynamic binding of functions would be a more attractive solution. These functions should not need to be defined in a special process control language defined by the supervisory system. Using a standard computer programming language instead means that the local embedded controller more easily can be integrated into quite different control systems.

In terms of PLC programming (IEC, 1992), we want to use a vendor independent method to change function blocks, or parts thereof, at run-time. Such a principle seems to be compatible with the evolving standard (IEC, 1992) for PLC and control system programming.

(The standard does allow use of externally defined functions and we have not found any requirement on static binding of such functions.)

2.3. Intelligent low-level sensing

Intelligent sensors are equipped with electronics and data processing capabilities so that data can be processed locally, thus requiring less data transfer and computations for the main control. Performance can also be improved by using intelligent sensors. Filtering algorithms in the sensor signal processing unit may run at a higher sampling rate, and operate on data that has not been delayed by data transfers.

High volume production of sensors like accelerometers and laser sensors have resulted in very low cost for the sensor. Field busses and interface electronics for connection of sensors are also becoming affordable. Such sensors today may provide a protocol for changing filter parameters and threshold levels. For a sensor to be considered intelligent, we also require that more general nonlinear and application specific algorithms can be performed in the sensor. There are, however, also parts of the filtering that are specific for the sensor technology, and should not be exposed to the user. This means that we want to have a hidden part of the filtering hardware/software, and an open part where application specific software should be possible to plug in. Such algorithms sometimes need to change over time, depending on the type of information needed by the high level control (like task planning, etc.).

As a simple test case, consider an accelerometer mounted on a robot arm. Normal robot motions result in smooth changes in acceleration. More rapid changes result if the robot hits an obstacle. We then want this event to be detected by the sensor and reported to the main control computer. If more information about the collision is requested, the sensor should also be able to reply with a recording of the signal during the collision. Such information can be provided by a special purpose intelligent sensor, or by an off-the-shelf intelligent **open** sensor.

3.   Embedded dynamic binding

We will now investigate how to bind functions at run-time to software already running in the embedded system, considering demands on efficiency and predictability for typical mechatronic applications.

3.1.   Alternatives

The problem we are facing can be described in terms of a client and a server. The server is the embedded software already implemented, and possibly running. The basic service is first of all to perform the control, but also to service requests to change the control behavior. It is the latter aspect that is treated here. The client acts on a higher hierarchical level, which means that it contains more application specific knowledge. Depending on what is appropriate to solve the particular application, the client requests the lower level control (i.e., the server) to use different new control functions. We then need some means of expressing these functions and to dynamically let the server evaluate them. As throughout this work, we try to apply well proven software techniques, but in the context of mechatronics and embedded control. Some of the alternatives we have considered for the implementation and dynamic binding are:

Remote Procedure Calls (RPC) are widely used and understood, but since the input/output arguments of the procedure have to be transferred to/from the computer that executes the procedure, the resulting data flows and communication delays can be substantial. Unpredictable timing would also be a problem, for instance when Ethernet connections are used.

Interpreted languages make it straightforward to pass functions to some target software containing an interpreter. Some languages to consider are:

Lisp is a flexible and powerful alternative, but tends to be one or two orders of magnitude slower than compiled languages unless the Lisp code is compiled. Its dependency on garbage collection also makes it hard to use for real-time tasks.

Forth is powerful and relatively efficient, but hard to debug and maintain due to its lack of structure. A solution useful for our needs could probably use Forth as an intermediate language, but with no apparent advantages.

Erlang (Armstrong, Virding and Williams, 1993) is an interpreted parallel language developed for telecommunication systems. It compiles to reasonably fast intermediate code that can be passes between processes running in different address spaces or on different CPUs, but as Lisp it relies on garbage collection, and we therefore decided against using it in our prototype implementation.

Java is a relatively new language with support for object-orientation, threads, and their relationship (Nilsen, 1996). It is compiled into efficient byte-code, but the drawbacks of Erlang remains.

Compiled functions written in a widely used language like C would be an attractive solution. The problem of binding the used symbols to the external environment then has to be solved. If this can be done, we will hopefully achieve almost the same flexibility as the other alternatives, and almost the same performance as for built-in functions.

Another drawback with the interpreted languages is that the system must always contain the (compiled) implementation of the interpreter, even when no additional software are used. The **Compiled functions** alternative, on the other hand, gives a wide range of language choices since almost any cross compiler can be used. In the sequel, the term *action* will be used for such a compiled function that is to be dynamically bound in the embedded system. This is the solution we have chosen for our project, and in the following sections we will investigate two different implementations of **actions**.

3.2. *Function-based actions*

Function-based actions are a chunk of code that has been compiled as position independent code (PIC). All interaction with the control system is done via an environment pointer that is passed as a parameter to the down-loaded function. The following implementation of a small example based on the application shown in Figure 1 further illustrates this approach. The header file environment.h contains the following C declaration of the environment:

```
typedef struct {
  float r, y, gain;        /* Names according to figure. */
  float (*sqrt)(float);
  /* more math functions here.. */
} Environment;
```

The following lines are then part of the standard implementation of the regulator:

```
#include <environment.h>

float defaultK(Environment *env) { return env->gain; }

float (*K)(Environment *) = defaultK; /* <- Action type. */

void Pcontrol() {
  Environment env;
  MakeDownloadable(K, "K");                /* <- Action slot. */
  env.sqrt = sqrt;
  for (;;) {
    env.r = GetRef();
    env.y = Sample();
    P = K(&env) * (env.r - env.y);    /* <- Control law. */
    SetOutput( P );
  }
}
```

where, as C-programmers know, the action declaration specifies that K is a pointer to a function that as its argument takes a pointer to the Environment specified above, and returns a float.

Since automatic code generation of control algorithms nowadays is an important control engineering tool, it is of course important not to impose special requirements on the way control laws are expressed. The fact that we use the structure env makes the control law (only the one line above in this simple example) different from the text-book version $P=K*(r-y)$ in two minor ways: First, accesses of variables that are part of the environment are preceded by 'env.'. That can for example be handled by preprocessor directives in C and C++, or by use of WITH in Modula-2. Secondly, the number K is represented by the function taking the address of the environment as an argument. (The "(&env)" can be avoided by using C++ operator overloading, but here we will be explicit and not confuse the reader by any syntactic sugar.)

Now the target system is prepared for a new function K to be down-loaded. Let us replace the defaultK with a separately compiled nonlinear gain function. Since the control law is the nonlinear gain times the control error, we have to divide the function g in Figure 1 by the control error. This means that the implementation of the action will be:

```
#include <environment.h>

float sqrtK(Environment *env) {
  float e = env->r - env->y;
  float a = 1 / (env->gain * env->gain);
  if      (e >  a) { return  1.0 / env->sqrt( e); }
  else if (e < -a) { return -1.0 / env->sqrt(-e); }
  else             { return env->gain; }
}
```

Further improvements concerning the control are possible, but outside the scope of this paper. Considering the software aspects the solution seems to be everything we need, but there are a few catches:

- The compiler must be capable to generate PIC code. On some platforms this can be prohibitively expensive in terms of execution speed or code size. It may also be the case that a certain compiler cannot generate PIC code.

- It is hard to make changes to the environment structure since all elements have to be in the same place, and no checks are done when the code gets down-loaded. Environment mismatches will result in run-time errors.

- Some language constructs need access to global data or global functions (e.g., operators **new** and **delete** in C++), which are not easily implemented in this scheme. This problem can be solved by a smarter compiler, but such a modification is probably not worthwhile.

In practice, this means that actions implemented this way are restricted to be single functions. This type of actions are therefore called function-based actions. Despite the problems, successful experiments have been carried out, as will be described in Section 4.

3.3. Dynamically linked actions

We will now introduce a linker to overcome the limitations with the function-based actions. Conceptually this seems more complicated, but in practice one can often use an existing linker and with minor effort generate a small file that contains the symbols that are needed to link with the control system. In our prototype system the linking is done with the GNU linker (**gld**) and a small Perl (Wall and Schwartz, 1991) script (300 lines) on the host system, and a few small procedures in the control system (400 lines + 1 line per symbol). The action client performs the following algorithm (by running the Perl script on the host computer):

1. The script requests all available interface symbols of a specific action slot from the control system.

2. A preliminary linking with these symbols is done to check that all references can be resolved. If any undefined symbols remain, linking is aborted.

3. Using the resulting code-size from step 2, the control system is instructed to allocate a chunk of memory with the appropriate size, and its start address is sent back to the client.

4. The final linking is done and the address of the entry point of the action is retrieved.

5. The finished code is sent to the control system along with the address of the entry point, see the function install below.

The implementation is straightforward. To make it possible to have multiple actions in the same code segment we will also change the way actions are installed in the example application. The following is the same example as above but somewhat modified, starting with the file environment.h:

```
typedef struct {
  float r, y, gain;
} Environment;

extern float sqrt(float);
/* more math functions here.. */

extern float (*K)(Environment *);
```

The following lines are part of the embedded control software:

```
#include <environment.h>

float defaultK(Environment *env) { return env->gain; }

float (*K)(Environment *) = defaultK;

void SendSymbols() {
  SendSymbol("sqrt", &sqrt);
  SendSymbol("K", &K);
}

void Pcontrol() {
  Environment env;
  MakeLoader(SendSymbols);
  for (;;) {
    env.r = GetRef();
    env.y = Sample();
    P = K(&env) * (env.r - env.y);   /* <- Control law. */
    SetOutput( P );
  }
}
```

As we can see, the things that have changed are:

- The **sqrt** function is no longer part of the environment structure, it is instead declared as an external function. By doing it this way, it is easy to add new functions to the interface, and the removal of old functions that are used by old actions will give diagnostic

117

messages when down-loading is attempted. In the C++ case we can also detect signature changes since the parameter types are encoded in the function names (name mangling). When Modula-2 is used, module time-stamp mismatches are detected in the same way.

- The action pointer is made visible to the down-loaded action. The reason is that the down-loaded action is now responsible for the installation, a fact that makes it possible to replace multiple actions simultaneously. This also makes it straightforward to add static C++ objects to a code segment.

Note that even if we could have placed all variables of the Environment outside the structure (as done with the math functions), the controller specific variables are kept in the Environment to allow several instances of a regulator to use the same action.

The clients implementation of the down-loadable action is slightly changed, and an install function is added.

```
#include <environment.h>

float sqrtK(Environment *env) {
  float e = env->r - env->y;
  float a = 1 / (env->gain * env->gain);
  if       (e >  a) { return 1.0 / sqrt(e); }
  else if (e < -a) { return -1.0 / sqrt(-e); }
  else              { return env->gain; }
}

void install() { K = sqrtK; }
```

In this case install() only connects the loaded action by assigning the address of the action to the action pointer. In more complex applications, install would probably take a number of arguments to be used for initialization of the loaded action segment. Such initialization may include calls of C++ constructors.

3.4. Experience

For each one of the two ways of implementing the dynamic binding, there are some advantages and drawbacks. First, the following benefits and problems with the **function-based** approach have been experienced:

+ Easy to understand and implement.

+ Code gets reentrant automatically, since all persistent data have to be kept in a pointer block. This means that the same code can be used at many places without any additional work.

+ Code blocks can be moved in memory after down-loading (compare Macintosh code resources (Apple, 1985)). This is a favorable property if memory compaction is needed.

− Global data can not be used.

- There are certain demands on the compiler; with GNU cc we had problems with non-optimized code and floating-point constants.

- On some architectures PIC-code is more expensive in terms of space and execution time.

- All calls to interface procedures have to be done by constant offsets into a jump-table; this makes it virtually impossible to remove anything from the interface (because old code would down-load correctly but use the old offsets which are incorrect for procedures located after the removed one).

- C++ devotees: Dynamic objects are hard to use as operator new and delete have to be global functions.

There are also some benefits and problems with the **dynamic linking** approach:

+ Any compiler can be used since all special work is done during linking.

+ Errors can be better detected when the interface is changed. In the C++ case signature changes are detected, and in the Modula-2 case time-stamp changes are detected.

+ The same programming model as in ordinary programs can be used. Global data and procedures are available, and even static C++ objects can be used.

- Special care has to be taken to write reentrant code. This is the same problem as in ordinary programs.

- The implementation in the target system is somewhat more complex.

- Code cannot be moved in memory since it is linked by the client to a specific address. This may lead to fragmentation problems.

   We will now discuss the use of these techniques on the applications that were presented in Section 2.

4. Experiments

The two principles presented in Section 3 have been fully implemented. Full implementations of the application examples have also been done. Some features of these implementations will be presented in this section. The presentation is hardware oriented.

4.1. Experimental environment

Implementation of the test-cases was carried out using an experimental robot control system that was developed for research in real-time systems, open robot control, and robot programming. The platform is built around commercially available robots. There is one ABB Irb-6 robot, and one ABB Irb-2000 robot which we will focus on in this paper. The control computers of the original system have been replaced by our VME-based control computer, connected to host computers via an Ethernet network. The VME system contains both

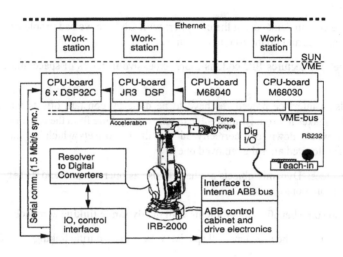

Figure 2. Overview of the experimental platform.

microprocessors and DSPs. This hardware configuration is typical for advanced embedded industrial motion control systems.

The sensors and actuators of the robot are accessed via an interface, which is connected directly to the serial ports of the DSPs. The hardware interfaces have been developed and built in our lab. The six joints of the robot can be controlled with a sampling rate of 8 kHz. A force sensor mounted on the robot end-effector is interfaced via another DSP board on the VME bus. An overview of the platform is shown in Figure 2.

4.2. Distributed systems

The use of actions to solve the supervisory control application exemplifies the benefits of actions in systems with distributed hardware connected via a computer network. An embedded control signal logging tool was developed in C++, and connected to the well known program Matlab running on the host workstation. A Matlab script was written and the graphics handling was tweaked for real-time performance (Figure 4 on Page 340 was created using this tool). The implementation showed that the use of actions is a good solution, but also the drawbacks mentioned in Section 3.4 were experienced. To avoid unnecessary details, we will now describe how the action concept can be conveniently handled in an object-oriented framework.

The client side communicates with the server that receives and executes the down-loaded actions. Under such circumstances, the supervisory control system running on a host workstation sends down actions to the server that executes the actions, thus computing the variables that are to be supervised. The software can be divided into the following five parts.

1. The client's compilation and linking strategy according to Section 3.

2. The transfer of the action from the client to the server. This part is standard computer communication and is therefore not further described.

3. The embedded software implementing allocation, initialization, storage, call, and deallocation of actions. This part corresponds to a generic holder of actions.

4. An interface part defining the environment (data and operations) that is available for actions. This defines the actual openness of the control object.

5. The specific application or control code executing the actions. Except for the initialization part, the control source code does not need to reflect that actions are allowed.

This means that we want to do an implementation for the embedded system of items 3, 4, and 5. An object oriented design and implementation using the C++ language was found to be well suited for this.

Figure 3 shows our object design. The class ActionLoader is independent of the way actions are transferred to the computer. This is instead encapsulated in the SockActLoader which uses some communication class, denoted Socket in the figure. The ActionSlot is a holder of a generic action. A number of type-safe classes derived from ActionSlot were written, one for each built-in type. These derived classes provide inline type-conversion operators for convenient use of actions in algorithms. For instance, the FloatAction contains an operator float.

4.3. Shared memory systems

Motion control systems often use multiple processors on a common bus. In less demanding applications when only one CPU working in one address space is used, a fixed linking and use of function pointers can provide the required flexibility. When using multiple CPUs that share memory via a bus, like the VME bus in Figure 2, we have to cope with different address spaces. Assume this is the case in our simple control example, and we want to use dynamically linked actions.

Using dynamically linked actions in a stand-alone system, as motion control systems often are, may seem to require the linker to be ported to the target architecture. Note, however, that the required functionality is very limited. It does not need to depend on file

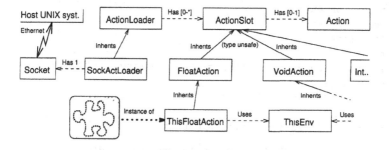

Figure 3. Classes for simple use of actions. The (type unsafe) inheritance from ActionSlot is part of the system, whereas the resulting action slot (ThisFloatAction in this case) is strongly typed.

systems and operating systems, command line decoding can be omitted, only a few options are needed, etc. Hence, such a linker is a procedure that is not very hard to write.

Source code for the simple control application has been presented in Section 3. The &env in the control law P=K(&env)*(r-y); is possible to omit by using overloaded operators in C++. In our case it was done by declaring K as a FloatAction (see Figure 3), which in C++ contains public: inline operator float();. This means that the action can be used wherever a float can be used, and the control law can be written as P=K*(r-y);, exactly as in C without actions. Standard tools for generation of control code can therefore still be used.

4.4. Dedicated hardware systems

The following implementation of the low-level sensing illustrates the use of actions for special purpose hardware. As shown in Figure 2, an accelerometer connected via an AD-converter to a DSP is available. For brevity, we will only use a simple action detecting an acceleration level. The following design tries to show how actions can be supported with a minimum of support from the embedded sensor software.

To reflect typical properties of intelligent sensors, our hardware was used as follows: 1) There is a memory area (called action_storage) reserved for loaded actions. 2) The action_storage is accessed by using the built-in DMA capabilities of DSP32C. This means that no support from the DSP program is needed; access of action_storage from outside only imposes some cycle stealing controlled by the hardware. 3) The address and the length of action_storage must be known by the action client. In our setup, we simply get this information from the symbol table of the linked sensor code. 4) The loaded action must obey any restrictions on the use of resources like CPU-time and memory. We used the DSP simulator (on the host computer) to check the execution time and use of memory.

We will now use function-based actions to catch collisions and plot the recorded signal in a host computer tool. Since they do not require any addresses of internal symbols (recall that the pointer to the environment is passed as a function argument at run-time), and the management of the action_storage is done by the master, the following C++ code is sufficient. First the header file included by both the action client (the master) and the action server (the sensor unit):

```
struct Environment {
  struct Signals {
    float time, acc;
  } *this_sample;
  enum status {pending, recording, transferring, waiting};
  status trig_state;
};


typedef int (*PlotAction)(Environment*);
```

As C-programmers know, the final typedef-line specifies that PlotAction is a pointer to a function that as its argument takes a pointer to the Environment specified above, and returns an integer. The basic sensor software then contains

```
int DefaultAction( register Environment *env ) {
  // Default behavior implemented here.
}
```

```
long action_storage[0x400];  // Size hard-coded for brevity.
PlotAction plot_action = DefaultAction;
```

The return value of the action tells whether sensor data output should be supplied or not. Our built-in data-supply object of type AccPlot contains the following member function:

```
int AccPlot::PutSample( float signals[] ) {
  // <Checking conditions that has to be fulfilled..>
  // <..16 lines omitted here..>
    // Space available, check trig condition.
    environment.this_sample = (Environment::Signals*)signals;
    if ((*plot_action)(&environment)) { // <- CALL OF ACTION.
      // Put data and let HW convert to IEEE output format.
      // <Some work done here.>
      return 1;
    } else {  // <Seven final lines omitted ..>
```

where the line containing (*plot_action)(&environment) shows how the action is called. As the reader with experience from the C language knows, this statement means that the function pointed to by plot_action is called, passing a pointer to the environment data structure as an argument. The following function was written and compiled, and then position independence, execution time, and memory usage was checked. This was done on the host workstation.

```
int AccPlotAction( register Environment *env ) {
  int ilevel = 25; float level;
  switch (env->trig_state) {
    case Environment::pending:
      level = float(ilevel);
      if (env->this_sample->acc > level) {
        env->trig_state = Environment::recording;
      } else { return 0; };
    case Environment::recording:  return 1;
    default:                      return 0;
  };
};
```

The threshold value, which in this example is fixed to $25m/s^2$, should in a real case get its value from some kind of parameter interface. The simple > test on the level can of course be replaced by more complex conditions, using for example filtering functions (provided via the Environment) from the application library of the DSP. The size of the compiled code for the AccPlotAction function is 212 bytes, and worst case execution time is $3.4\mu s$. It was loaded into the DSP memory using DMA, and it was activated by assigning the address

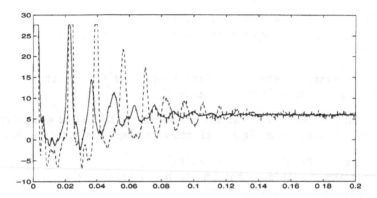

Figure 4. Acceleration $[m/s^2]$ versus time $[s]$ after contact event. The solid line shows the desired response with the gasket applied, and the dashed line shows the metal contact when the gasket is missing.

of the loaded action to the plot_action variable (done as one atomic operation using the DMA facility). With a sampling rate of 8 kHz, a recording of the transient when a slow downward robot motion touches a cover plate made of metal was requested. For proper assembly, there should be a gasket applied on the plate. The intelligent sensor catches the contact transient and sends the captured data to the host computer. The rigidity/compliance of the surface affects the response quite significantly as shown in Figure 4. It is therefore a trivial task for the high level control to decide if the gasket is missing or not.

There are, however, also problems with the above solution. As the observant reader has noticed, the threshold level in the AccPlotAction function is first given an integer value 25, which is then converted to a float. It would of course be better to declare floating point values directly, but the standard DSP32C C-compiler does then not produce position independent code. This is typical for RISC architectures, which uses a fixed instruction length. In this case, the instruction length is 32 bits. That is also the length of a float (and also of a long), whereas an integer occupies 24 bits. The C-compiler therefore includes the constant integer values in the instructions in the code-section. A float, on the other hand, will be put in the data section, thus requiring linkage to the executable code. The same problem occurs, as mentioned in Section 3, when auxiliary functions are included in the action.

Even if the problem is due to the compiler, it is still relevant since compilers generally work in this way, and actions are preferably written in a high-level language. Special languages and/or compilers are sometimes used for safety critical systems to get deterministic execution properties. Such a special compiler could provide position independence also for the above function. It is also easy to achieve that using assembly language. Despite the problems, we therefore think that function-based actions can be a useful alternative in some applications. But for portability and execution efficiency reasons, the dynamic linking version is probably the best alternative in most cases.

5. Safety and predictable real-time performance

The proposed mechanism is a powerful way to add flexibility to embedded systems, but how are safety and predictability affected? The following investigation is made with some typical properties of hard real-time systems in mind. The reader may think of the Spring Kernel (Stankovic and Ramamritham, 1988) and the Synchronous approach (Simon et al., 1993) as relevant examples.

Since we have the same requirements on safe programming of actions as for implementation of the standard part of the system, some way to ensure that safety has to be devised. If a language such as C is used to implement the actions, the system has to protect itself from invalid memory accesses that otherwise could compromise the entire system. Proper programming of the MMU can prevent illegal memory accesses but that is an expensive solution, especially concerning worst case execution time for dynamically linked actions. A more convenient (for the programmer) and efficient (for the computer) approach is to use a language (like Java (Nilsen, 1996)) where memory accesses can be checked a priori.

To ensure that deadlines in the system are met, the timing of individual actions is of paramount interest. If the system is statically scheduled, the schedule has to take into account the extra time actions are allowed to take. The timing constraints are then a property of the action slot, and the constraints have to be checked when actions are installed. In dynamically scheduled systems, rescheduling has to be done when actions are loaded. In both cases, the execution time of the action has to be determined.

In some cases, as in our DSP system, execution times can be determined by the client, and the action loader can check that timing constraints are fulfilled. Thus, no additional functionality/software for timing analysis is needed on the target side. In other more general cases, for instance when the action may be down-loaded to differently configured hardware (but with compatible instruction sets), it is more suitable to perform timing evaluation of the action in the target system. To determine the actual timing of the action, it either has to be statically analyzable, like in Real-Time Java (Nilsen, 1996), or the timing has to be checked at run-time by measuring the actual time taken for some appropriately chosen test runs of the action. Assume that each action contains a function that calls the loaded function with special parameters to get the maximum execution time. It is not appropriate to just let the `install` function (see the code example on Page 334) call the `MaxTime` function and then compare with the bounds by calling some evaluation function defined in the action's environment. This is because timing analysis may also require the hardware (interrupts, caches, etc.) to be setup in the same way as when the action has been installed. Therefore, we propose that action slots in statically scheduled systems require actions to have a `MaxTime` entry (and `MinTime`, etc.), and that the server calls it before the `install` function. This is a minor and straightforward extension of our present implementation. Of course, the program must have a properly scheduled time slot for the timing evaluation, but that slot can serve all actions managed by that CPU.

When new code has been properly installed, it can be useful to monitor signals emanating from the the action to make sure that it fulfills predefined control constraints. This was proposed in the Simplex Architecture (Sha, Rajkumar and Gagliardi, 1995), which has been developed to support online upgrade of hardware and software components. The action mechanism also facilitates such a feature; actions can of course be removed (fully implemented but omitted here) when conditions of activation are not fulfilled.

6. Conclusions

A development towards more open computer systems is motivated by increased demands on flexibility, i.e., systems should be possible to tailor to meet new customer demands. The need for open systems also within mechatronics and embedded control was illustrated by some application examples. To permit feedback control based on information from external sensors, and for the design of reactive systems in general, open control systems are needed. Special demands on efficiency, safety, etc. motivated use of dedicated software techniques for this type of systems.

The most basic way to allow changes is by providing parameters that can be tuned by the user. Such parameters have been restricted to data, but the demands on flexibility often require the program to be changed. Our idea is to use pieces of compiled executable code as functional operators. Such operators may in the simplest case be a linear function, which combined with operators provided by the programming language may appear as an ordinary numerical control parameter. In more complex situations, even if encountered after implementation of the core system, we may introduce a function with additional arguments (states, time, etc.), or even a 'function' with internal states. A new parameterization of the control is then obtained. In an object-oriented framework, this means that new methods can be added to controller objects after implementation of the basic control, and even while the controller is running. Two ways to do the required dynamic binding of functions and objects were introduced and discussed.

A major effort was put into full implementation of the test cases. The results were encouraging; it was possible to achieve flexibility and efficiency. In other words, an embedded system may indeed be open, thus justifying the title of this paper. The application examples, the proposed principles for open embedded control, and the implementations verifying the performance are the main contributions.

In conclusion, we propose a way to build **open and efficient** embedded systems. It was also described how other research results, concerning for example predictability and static scheduling (Stankovic and Ramamritham, 1988; Stankovic and Ramamritham, 1993), can benefit from the proposed mechanism. We think that a major advantage of the concept is its simplicity and the possibility to combine it with other principles within mechatronics and real-time systems.

Acknowledgments

NUTEK – the Swedish National Board for Industrial and Technical Development – Embedded Systems Research Program – Grant 93-02726.

References

Apple Computer Inc. 1985. *Inside Macintosh, I.* Addison-Wesley.

Apple Computer Inc. 1994. *OpenDoc for Macintosh: An Overview for developers.*

Armstrong, J., Virding, R. and Williams, M. 1993. *Concurrent Programming in Erlang.* Prentice Hall.

Åström, K. and Hägglund, T. 1995. *PID Controllers: Theory, Design, and Tuning, second edition.* Instrument Society of America, Research Triangle Park, NC.

Dorward, J. and Sethi, R. 1990. Adding new code to a running C++ program. *USENIX C++ conference proceedings,* pp. 279–292.

Ford, W. 1994. What is an open architecture controller. *9th International Symposium on Intelligent Control.*

IBM Corp. 1993. An introductory guide to the System Object Model and its accompanying frameworks. *SO-Mobjects Developer Toolkit, Users Guide, version 2.0.*

International Electrotechnical Commission. 1992. *IEC 1131-3, Programmable controllers, Part 3: Programming languages.*

Johannesson, G. 1994. *Object-Oriented Process Automation with SattLine.* Chartwell Bratt Ltd.

Microsoft Corporation. 1993. *Object Linking Embedding, version 2.0.*

Nilsen, K. 1996. Real-Time Java. Technical Report, Iowa State University, Ames, Iowa.

Nilsson, K. and Nielsen, L. 1993. On the programming and control of industrial robots. *International Workshop on Mechatronical Computer Systems for Perception and Action,* pp. 347–357, Halmstad, Sweden.

Object Management Group (Digital, HP, HyperDesk, NCR, Object Design, SunSoft). 1991. *The Common Object Request Broker: Architecture and Specification,* document number 91.12.1 edition.

Sha, L., Rajkumar, R. and Gagliardi, M. 1995. A software architecture for dependable and evolvable industrial computing systems. Technical Report CMU/SEI-95-TR-005, Software Engineering Institute, CMU, Pittsburg.

Simon, D., Espiau, B., Castillo, E. and Kapellos, K. 1993. Computer-aided design of a generic robot controller handling reactivity and real-time control issues. *IEEE Transactions on Control Systems Technology,* pp. 213–229.

Stankovic, J. and Ramamritham, K. 1988. *Hard Real-Time Systems.* IEEE Computer Society Press.

Stankovic, J. and Ramamritham, K. 1993. *Advances in Real-Time Systems.* IEEE Computer Society Press.

Sun Microsystems. 1993. *SunOS Linker and Libraries Manual.*

Wall, L. and Schwartz, R. 1991. *Programming Perl.* UNIX Programming, O'Reilly & Associates.

Real-Time Systems, 14, 345–351 (1998)
© 1998 Kluwer Academic Publishers, Boston.

Contributing Authors

Paolo Ancilotti received the Electronic Engineering degree from the University of Pisa, and the Diploma of the Scuola Superiore S. Anna, both in 1967. He worked for the National Research Council at the Institute for Information Processing in Pisa and then he has been on faculty with the Department of Information Engineering at the University of Pisa. Now he is full professor of Computer Engineering and Dean of the Faculty of Applied Sciences at the Scuola Superiore S. Anna in Pisa. His research interests include Operating Systems, Concurrent Programming, Software Engineering and Real-Time Systems.

Anders Blomdell is a research engineer at Department of Automatic Control, Lund Institute of Technology (or Lund University), Sweden. His professional interests cover real-time programming, man-machine interaction, computer language design and networking protocols. He is a strong proponent of strongly typed languages, programming by contract, real-time garbage collection and object-organization, but is still looking for a real-time language that has proven to be better than Modula-2 (his current opinion is that Java is a promising candidate).

129

Giorgio C. Buttazzo is an assistant professor of Computer Engineering at the Scuola Superiore S. Anna of Pisa, Italy. He received a B.E. degree in Electronic Engineering at the University of Pisa in 1985, a M.S. degree in Computer Science at the University of Pennsylvania in 1987, and a Ph.D. degree in Computer Engineering at the Scuola Superiore S. Anna of Pisa in 1991. During 1987, he also worked on active perception and real-time control at the G.R.A.S.P. Laboratory of the University of Pennsylvania. His main research interests include real time computing, dynamic scheduling, multimedia systems, advanced robotics, and neural networks. Dr. Buttazzo has authored, and co-authored, three books on real-time systems and over sixty journal and conference papers.

Matjaž Colnarič is an associate professor at the Faculty of Electrical Engineering and Computer Science, University of Maribor, Slovenia, from which he also received his Master of Science and Doctor of Science degrees in 1983 and 1992, respectively. He teaches courses on microprocessors, real-time systems, and algorithms and data structures. His main research interests are related to embedded real-time control systems, their hardware and system architectures, operating systems, programming languages as well as application design techniques and methodologies for them. Dr. Colnarič has authored, or co-authored, about fifty journal and conference papers, mostly in the real-time area. He is a member of the IEEE Computer Society and its TCs on Real-Time Systems and Complexity in Computing. He is also a member of IFAC Technical Committee on Real-Time Software Engineering.

Marco Di Natale is an Assistant Professor at the University of Pisa, Italy, at the Information Engineering Department. He received a Ph.D. degree in Computer Science from the Scuola Superiore S. Anna, Pisa, Italy in 1995 and in 1994 was part of the Spring group at the University of Massachusetts in Amherst. He is currently cooperating with the Retis Lab. at the Scuola S. Anna doing research on real-time systems. His main research interests include scheduling algorithms, real-time communication protocols, design methodologies and tools for the development of real-time systems.

Manfred Glesner was born in Saarlouis (Germany) and got his university education at the Saarland University in Saarbruecken, Germany. He studied Applied Physics and Electrical Engineeering and got his diploma in 1969. In 1975 he received from the same university the Ph.D.-degree for his research work on optimization techniques in computer aided circuit design. In 1981 he became Assistant Professor in the Faculty for Electrical Engineering and Information Technology of the Technical University of Darmstadt (=Technische Hochschule Darmstadt). Since 1989
he owns the chair for Microelectronic Systems at Darmstadt University. In education he is responsible for teaching courses in microelectronic circuit design and courses in CAD for microelectronics. In research he is interested in topics of high-level synthesis, physical design especially for deep submicron technologies and VLSI for digital signal processing. Further he is responsible in a Special Research Area of the German National Science Foundation (=DFG) for microelectronic system design in mechatronics. He is a member of several national and international professional organizations. He organized several national and international conferences, and he is a member of the programme committee of many conferences and workshops. For his scientific contributions to the field of microelectronics he received two honorary causa doctoral degrees from Tallin Technical University (Estonia) and from Bucharest Polytechnical University (Romania).

Roman Gumzej joined the Faculty of Electrical Engineering and Computer Science at University of Maribor, Slovenia, in 1994, as a junior researcher. He received his Master of Science degree in 1997, and is currently working towards a doctorate. His main research interests are real-time operating systems, and verification and validation of real-time systems.

Wolfgang A. Halang, born 1951 in Essen, Germany, received a doctorate in mathematics from Ruhr-Universität Bochum in 1976, and a second one in computer science from Universität Dortmund in 1980. He worked both in industry (Coca-Cola GmbH and Bayer AG) and academia (University of Petroleum & Minerals, Saudi Arabia, and University of Illinois at Urbana-Champaign), before he was appointed to the Chair for Applications-Oriented Computing Science and head of the Department of Computing Science at the University of Groningen in the Netherlands. Since 1992
he holds the Chair for Computer Engineering in the Faculty of Electrical Engineering at FernUniversität Hagen in Germany. His research interests comprise all major areas

of hard real time systems with special emphasis on safety licensing. He is the founder and European editor-in-chief of "Real-Time Systems," member of the editorial boards of 4 further journals, co-director of the 1992 NATO Advanced Study Institute on Real Time Computing, has authored 7 books and some 240 refereed book chapters, journal publications and conference contributions, has given some 60 guest lectures, and is active in various professional organisations and technical committees as well as involved in the programme committees of numerous conferences.

H.-J. Herpel received the M.S. and Ph.D. degrees in electrical engineering from Darmstadt University of Technology in 1985 and 1995, respectively. From 1985 to 1988 he worked as system engineer for Dornier Systeme GmbH. He was responsible for the specification and design of onboard software for various space programs, e.g., the environmental control and life support system of the european space shuttle HERMES. In 1989 went back to Darmstadt University of Technology to work as research assistant in a mechatronics project. 1994 he left Darmstadt University for a project manager position in the microsystems technology programme of the Forschungszentrum Karlsruhe GmbH. One year later he entered Daimler-Benz Aerospace in Friedrichshafen to manage simulation projects in the field of spaceborne communication and navigation systems.

Olof Laurin received his MS degree in Electrical Engineering from Lund Institute of Technology, Sweden, in 1995. His main topics were real-time systems and control theory, and in his masters project he worked with dynamic binding of executable code, including object-oriented support for that. Olof is now working at ADRA Datasystem AB in Malmö, Sweden. For the last year he has been working mainly with database design.

Klas Nilsson received his MS degree in Mechanical Engineering from Lund Institute of Technology, Sweden, in 1982. He then worked six years at ABB Robotics Products in Västerås, Sweden, where he developed motion control software and algorithms. Having experienced the contradiction between open/flexible systems (needed for new advanced applications) and optimized/low-cost products, Dr. Nilsson started his PhD studies at the Dept. of Automatic Control in Lund, Sweden. His thesis, entitled "Industrial Robot Programming" (1996), pays careful attention to the interaction between programming and control of robots, including dynamic management of executable code to improve robot applicability. During parts of 1996/97 he returned to ABB Robotics and participated in the development of future robot programming principles. Doing research and teaching in robotics and real-time systems, he currently works as an Assistant Professor at the Dept. of Computer Science in Lund, Sweden.

Marco Spuri has recently held a visiting research position at INRIA, Rocquencourt, France, where he worked within the REFLECS project on real-time distributed systems. His research interests include real-time computing, operating systems, and distributed systems. He graduated in Computer Science at the University of Pisa, Italy, in 1990. In the same year, he also received the diploma of the Scuola Normale Superiore of Pisa. In 1992 he joined the Scuola Superiore S. Anna of Pisa, where he received a doctorate degree in Computer Science in 1995.

Hans Thielemans, born at Mechelen, Belgium on January 8, 1957, obtained the degree in mechanical engineering from the Katholieke Universiteit Leuven, Leuven Belgium. He is now a senior research associate of the division Production Engineering, Machine Design and Automation of the Mechanical Department at the Katholieke Universiteit Leuven. He has been active in the research on robot compliant motion control with active compliance, space robotics, machine tool control, servo control systems and real-time computer systems. At the moment he is working in several space related robotic projects and in an ACTS projects on teleoperation. He is project manager for a Brite project on modular motion control systems with emphasis on synchronisation aspects.

Martin Törngren is a research associate at the Mechatronics lab at the Royal Institute of Technology (KTH) in Stockholm. He is active in both research and education in the area of real-time control systems. His research interests cover several aspects of control system design and implementation, including computer aided tool support, decentralization, real-time software architectures and sampled data-control systems (multirate and time-varying systems). Dr. Törngren is currently developing and running postgraduate and undergraduate courses in real-time control systems, one of which is currently made available for interactive education via WWW.

Hendrik Van Brussel, born at Ieper, Belgium on October 24, 1944, obtained the degree of technical engineer in mechanical engineering from the Hoger Technisch Instituut in Ostend, Belgium in 1965 and an engineering degree in electrical engineering at M. Sc. level from Katholieke Universiteit Leuven, Belgium. In 1971 he earned a Ph. D. degree in mechanical engineering, also from K.U.Leuven. From 1971 until 1973 he was establishing a Metal Industries Development Centre in Bandung, Indonesia where he also was an associate professor at Institut Teknologi Bandung, Indonesia. Subsequently, he returned to K.U.Leuven where he is presently full professor in mechatronics and industrial automation and chairman of the department of mechanical engineering. He was a pioneer in robotics research in Europe and an active promoter of the mechatronics idea as a new paradigm for concurrent machine design. He has published more than 200 papers on different aspects of robotics, mechatronics and flexible automation. His present research interests are shifting towards holonic manufacturing systems and precision engineering, including microrobotics. He is Fellow of SME and IEEE and in 1994 he received an honorary doctor degree from the 'Politehnica' University in Bucarest, Romania and from RWTH, Aachen, Germany. He is also a corresponding member of the Royal Academy of Sciences, Literature and Fine Arts of Belgium and an active member of CIRP (International Institution for Production Engineering Research).

Domen Verber is a teaching assistant at the Faculty of Electrical Engineering and Computer Science, University of Maribor, Slovenia, from which he received his Master of Science degree in 1996. He is currently working towards his doctoral degree. His main research interests are real-time programming languages, object orientation, temporal and schedulability analysis of tasks, and application design techniques and methodologies. He is a member of the IEEE Computer Society.